Domestic Food Production and Food Security in the Caribbean

Domestic Food Production and Food Security in the Caribbean

Building Capacity and Strengthening Local Food Production Systems

Clinton L. Beckford and Donovan R. Campbell

Softcover reprint of the hardcover 1st edition 2013 978-1-137-29698-6

First published in 2013 by
PALGRAVE MACMILLAN®
in the United States—a division of St. Martin's Press LLC,
175 Fifth Avenue, New York, NY 10010.

Where this book is distributed in the UK, Europe and the rest of the World,
this is by Palgrave Macmillan, a division of Macmillan Publishers Limited,
registered in England, company number 785998, of Houndmills,
Basingstoke, Hampshire RG21 6XS.

Palgrave Macmillan is the global academic imprint of the above
companies and has companies and representatives throughout the world.

Palgrave® and Macmillan® are registered trademarks in the United
States, the United Kingdom, Europe and other countries.

ISBN 978-1-349-45197-5   ISBN 978-1-137-29699-3 (eBook)
DOI 10.1057/9781137296993

Library of Congress Cataloging-in-Publication Data

Beckford, Clinton L.
   Domestic food production and food security in the Caribbean :
   building capacity and strengthening local food production systems /
   Clinton L. Beckford and Donovan R. Campbell.
      p.   cm.
   Includes bibliographical references.

   1. Food supply—Caribbean Area.   2. Food security—Caribbean
   Area.   3. Farms, Small—Caribbean Area.   I. Campbell, Donovan R.
   II. Title.
   HD9014.C442B43 2013
   338.1'9729—dc23   2012050008

A catalogue record of the book is available from the British Library.

Design by Integra Software Services

First edition: June 2013

10 9 8 7 6 5 4 3 2 1

Transferred to Digital Printing in 2013

Contents

Part III Environmental Change, Constraints, Vulnerability, and Survival in Small-Scale Food Farming Systems in the Caribbean

Part IV Globalization and Caribbean Food Security: Challenges and Opportunities

Part V Synthesis

List of Illustrations

Figures

Tables

Boxes

Preface

Defining the Caribbean can be as intriguing as the countries of the region themselves. The term *Caribbean Basin* is sometimes used to capture or encompass all the countries commonly included in definitions of the region. Potter et al. (2004) describe the Caribbean Basin as having three geographic groupings—the Greater Antilles, the Lesser Antilles, and the Islands of the Bahamas and Turk and Caicos archipelagos. A mention of the Caribbean conjures up mental images of islands, and the region is indeed largely made up of islands and island chains. However, as Potter et al. point out, the north coast of South America fringes the region and the countries of Guyana and Suriname are considered to be parts of the Caribbean, even though geographically they are part of the South American continent. The inclusion of the mainland countries of Guyana and Suriname in the Caribbean can be attributed to their shared history and culture with the West Indies. Economic relations with another mainland country, Belize, have led to its inclusion in the CARICOM group.

Potter et al. (2004) aptly describe the Caribbean as a "diverse, vibrant and complex world region" (p. XVIX). The Caribbean countries reflect a rich colonial past, with ties to England, Spain, France, and the Netherlands. Through the imperialistic goals of the colonists, ties were developed with both Africa and Asia, which supplied the region with labor in the form of slaves from Africa, and after slavery ended in 1838, with indentured servants mainly from what was then known as the East Indies. The Caribbean region holds considerable intrigue internationally. The natural beauty of the countries is no doubt a big reason for this, but the fascinating history of the region and its reggae and calypso music, world famous cuisine, and prowess in sport are also contributing factors. Since the first instances of conquest and colonization in the Caribbean, the region has been a major player in international politics and economics. In the slave era and the longer colonial period, the economies of the region were tied to metropolitan centers in Europe. The countries used to produce primary products for the European countries, while they had to import everything they needed, including food. The legacy of this mercantilist era is evident in the region today, with great implications for regional food security.

This book focuses on the countries of the region that constitute the Caribbean Community, or CARICOM. The CARICOM is a political and economic organization essentially organized around efforts toward a common market and toward scale economy in this context. The CARICOM member countries are depicted in Table P.1. Reference is also made to CARIFORUM—a political unit around which the region's food security strategy is based. The CARIFORUM is CARICOM plus the Dominican Republic.

In addition, there are a number of countries that are associate members—Anguilla, Bermuda, British Virgin Islands, the Cayman Islands, and Turks and Caicos Islands. The Caribbean Basin is depicted in figure P.1; the figure also highlights the CARICOM countries.

Like all regions, the Caribbean, including the CARICOM countries, is a bit of an enigma in terms of its characteristics. Homogenity is a key feature, but diversity is also another characteristic of the region, and all the countries have their own unique quality. The region is diverse in terms of size of land mass, population size, language and culture, ethnic composition, and socioeconomic development. The Caribbean countries are similar in that they are mostly small island developing states (SIDS), which have recently gained independence from Britain. Guyana, Belize, and Suriname are mainland countries; Haiti and Belize were not British slave colonies, and their population speaks French and Spanish, respectively. In addition, Haiti is neither a mainland country nor an island but is one of two countries on an island that used to be known as Hispaniola. Influence of the Netherlands is manifested in the mainland country of Suriname, which is a border neighbor of Guyana. Many of the countries of the region have changed hands throughout their histories and many reflect multiple European influences.

The CARICOM countries generally have high rates of adult literacy, high rates of emigration, and low birth rates, leading to low population growth rates; they have young populations, high unemployment rates, and high rates of HIV/AIDS (Caribbean Food and Nutrition Institute/Food and Agricultural Organization, 2007). Nine of the 13 countries have predominantly rural populations, yet on aggregate, the urban population in the region exceeds the rural population. It could be argued that for a region of its size, the CARICOM is rich in resources including bauxite, petroleum, natural gas, forestry, and precious metals. It also boasts an internationally famous tourism product dominated by Jamaica, the Bahamas, Barbados, St. Lucia,

Table P.1 CARICOM member countries

Countries	Political status	Land area (km²)	Population 2012	GDP (Nominal)	GDP (PPP) (US$) Millions	GDP Per capita (US$) Millions	Human Development Index
CARICOM			**16, 926, 054**		**103, 235**	**13, 888**	**0.558** (incl. Haiti) **0.723** (excluding Haiti)
Antigua/Barbuda	Independent	440	89,018	1,425	1,781	21,363	0.764
Bahamas	Independent	13,900	316,182	7,700	10,786	30,009	0.771
Barbados	Independent	431	287,733	4,110	6,300	22,744	0.793
Belize	Independent	22,970	327,719	1,401	2,804	8,069	0.699
Dominica	Independent	750	73,126	476	981	13,222	0.724
Grenada	Independent	340	109,011	789	1,396	13,135	0.748
Guyana	Independent	215,000	741,908	2,258	5,842	7,038	0.633
Haiti	Independent	27,750	9,889,187	6575	12,421	1,167	0.454
Jamaica	Independent	10,900	2,889,189	13,356	24,737	8,747	0.727
St. Kitts/Nevis	Independent	260	50,726	676	925	16,218	0.735
St. Lucia	Independent	620	162,178	1,198	2,158	12,476	0.723
St. Vincent/Grenadines	Independent	390	103,537	684	1,259	11,561	0.717
Suriname	Independent	163,000	560,157	3,682	5,069	8,949	0.680
Trinidad/Tobago	Independent	5,130	1,326,383	20,375	26,866	19,739	0.760

Sources: World Bank (2006); GDP data-IMF (2011); HDI data—UNDP (2012); Population data—CIA World Factbook, (2012).

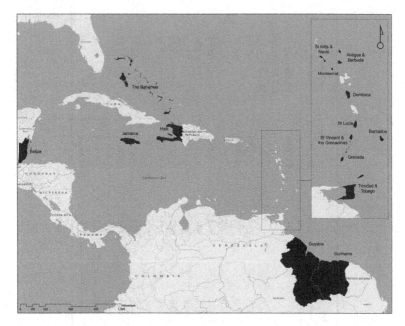

Figure P.1 Map of the Caribbean depicting CARICOM member countries

Antigua/Barbuda, and St. Vincent and the Grenadines. In every country of the region, agriculture's contribution to the GDP and development is declining, but it still remains an important part of the region's economy. Only in Haiti is agriculture the main contributor to the GDP—everywhere else, services and manufacturing have become the dominant drivers of growth.

Agriculture in the CARICOM has been influenced by the history of the region. The state of food security in the region today is tied to its history and past priorities. The region developed as a plantation economy tied to international markets by a mercantilist policy in which agricultural production was geared toward export to Europe, and food for the colonies was imported. The neglect of domestic agriculture historically is still a characteristic of the contemporary Caribbean, and according to some sources, this has resulted in the region becoming a net importer of food, with increasing food insecurity (Timms, 2008). With the exception of Haiti, the sensationalized issues of hunger reported in certain parts of the developing world are largely unknown in the Caribbean. The region has experienced rising caloric availability since the 1960s, well above recommended amounts. Food deprivation has decreased regionally, but still over 6 million people were considered food deprived in 2003 (CFNI/FAO, 2007). The CFNI reports that fruit and vegetable availability has increased sharply in the region, but is still lower than recommended values. Despite this, there are growing concerns about the state of food security in the region, with declining domestic production, increasing nutritional diseases, and increased dependence on imported food, causing vulnerability in food security. There is also a growing incidence of nutrition-related diseases (CFNI/FAO, 2007).

The region has become highly dependent on imported food, with a rising food bill and growing trade imbalance (Deep Ford and Rawlins, 2007). Currently, only Belize and Guyana produce more food than they import—the other countries in the region are net importers of food. Food imports "have crowded out regional production and the region's competitive disadvantage vis a vis imports has led to losses in employment, food security and rural well-being" (Barbados Advocate, 2012). The CARICOM Secretariat estimated the region's food import bill (excluding Haiti) in 2008 at some US$ 3.5 billion (FAO, 2008). Although caloric availability is up, food insecurity is still an issue. The region has developed a taste for North American diets and has become estranged from traditional foods. The use of fats and sugars is excessive, and nutrition-related diseases are on the rise (CFNI/FAO, 2007). Spurred by obesity, which is a fairly recent and growing problem, diabetes, high blood pressure, heart attack and stroke, and some forms of cancer are prevalent and are major causes of death. Pockets of persistent poverty have compromised the ability of many families to eat enough and eat well. The percentage of people living below the poverty line ranges from a low of 8 percent in Barbados to an alarming 74 percent in Haiti. An analysis of the food security status of the region would conclude that food availability is not a major issue. Rather it is the lack of access to food caused by poverty and food consumption patterns that constitutes the main threat to food security (CFNI/FAO, 2007). We would add the dependence on imported food as a major concern that, from a livelihood perspective, victimizes local smallholder farmers and rural families and makes regional food security vulnerable.

This book examines some of the contemporary issues impacting food production and food security in the CARICOM countries of the Caribbean. Using a livelihood conceptual framework, the book advocates the enhancement of domestic food production as the most appropriate way to improve the overall food security and discusses strategies for building capacity in local food production systems.

We would highlight a few features of the book that emphasize its value at this time. First, this book looks at the topical and pressing issue of food security, which is a global concern. Second, it studies an underrepresented geographic region—the Caribbean. Third, we have attempted to write in a way that will appeal to a wide cross-section of audiences including academics and undergraduate and graduate students. Also the book is the result of empirical research and syntheses of research conducted in the region, much of it by Caribbean scholars and much of it by researchers from the Department of Geography of the University of the West Indies. There is a fairly heavy dose of Jamaica in the book, which reflects the source of most of the available literature and research. We have drawn heavily on our own individual and collaborative research, which is based in Jamaica, but we synthesize the research of other scholars to provide a truly Caribbean perspective. We are confident that this book will appeal to a wide international multidisciplinary audience.

We have sought to structure the volume in a way that, we hope, readers will find logical, thematic, and easy to follow. The book is divided into five sections based on broad themes, but we recognize that these are not necessarily mutually exclusive:

1. Foundations of Food and Agriculture in the Caribbean
2. Salient Issues in Food Production and Food Security in the Caribbean
3. Environmental Change, Constraints, Vulnerability, and Survival in Small-Scale Food Farming Systems in the Caribbean
4. Globalization and Caribbean Food Security: Challenges and Opportunities
5. Synthesis

The book is a product of over ten years of research by the authors, and its targeted audience includes scholars and students of Caribbean geography, cultural geography, food and agricultural geography, and food security. We were inspired to write the book because of the lack of a comprehensive synthesis of food and agriculture in the Caribbean. We have worked with small-scale farmers in Jamaica over the years, whose resilience and steadfast goal of producing food in a challenging environment have impressed us. Research from across the region has shown that this is a characteristic of small-scale farming systems throughout the region. This book, therefore, pays homage to these farmers and calls for a greater institutional support for their efforts.

PART I

Foundations of Food and Agriculture in the Caribbean

CHAPTER 1

The Role of Agriculture in Caribbean Economies: A Historical and Contemporary Analysis

1.1 Overview

Agriculture has always been integral to the economies of the Caribbean. Potter et al. (2004) posit that the evolution of regional societies and economies in the Caribbean has been fundamentally determined by the plantation era and the inequalities and inequities created by slavery and colonialism. After emancipation in 1838, "new rural land use patterns and agrarian structures emerged reflecting a different, though still unjust social order" (Potter et al., 2004, p. 97). In the Caribbean of the post-emancipation period, there emerged a local peasantry made up of ex-slaves who left the sugar plantations and established independent communities, called free villages, with economies based on small-scale agriculture and other informal activities such as small-scale retailing, fishing, and charcoal burning. With export agriculture based on sugar still dominant, there developed an agricultural system characterized by a structural dualism (Barker, 1989, 1993). The features of this system are a large-scale export-oriented sector based on traditional plantation crops like sugarcane and banana juxtaposed with a small-scale farming sector focusing on domestic food crops that are staples in local diets and local cuisine (Barker, 1993; Beckford and Bailey, 2009). This duality has persisted and characterizes agriculture in many Caribbean countries to this day.

Agriculture in the Caribbean is characterized by significant diversity, with each island having its unique landscape shaped by economic imperatives, cultural history, and environmental realities (Potter et al., 2004). Agriculture benefits Caribbean economies in a number of ways. These include contribution to gross domestic product (GDP), export earnings, employment, and domestic food supply. It provides industrial raw materials, is the basis of rural livelihoods, and is still a major land-use activity in many of the countries in the region (Deep Ford and Rawlins, 2007; Potter et al., 2004).

Agriculture is still very important to the Caribbean but its economic significance is changing as economic imperatives shift and island economies diversify to become more industrial based and service and technology oriented. In terms of agriculture's contribution to GDP, there is a declining regional trend (see table 1.1).

Regionally, agriculture makes up around 25 percent of GDP, making the highest contributions to the economies of Haiti and Guyana. This compares to around 5 percent in developing countries and above 40 percent in Africa and Asia. In St. Kitts and Nevis, agriculture's contribution to GDP fell from 40 percent in 1964 to less than 6 percent in 2000. In Barbados, it declined from 38 percent in 1958 to 4 percent in 2000 (Potter et al., 2004). Unlike the developed world, where agriculture's declining contribution to GDP coincided with a rapid growth in urban industrialization, in the Caribbean the decline is matched by a rapid expansion of the service sector, which is dominated by tourism. Regionally tourism now contributes up to 75 percent of GDP, with notable exceptions being Guyana and Haiti, where its contribution is less than 50 percent (Potter et al., 2004). Ironically the

Table 1.1 Comparative contribution to GDP in the CARICOM

Countries	Agriculture (percent)	Industry (percent)	Services (percent)
Antigua/Barbuda	3.8	32.9	51.8
Bahamas	1.6	8	90.4
Barbados	3.4	13.4	83.2
Belize	9.7	19.8	58.
Dominica	21.4	22.4	56.3
Grenada	10.6	28.9	60.5
Guyana	24.2	24	51.8
Haiti	25	16	59 (2010 estimate)
Jamaica	5.8	29.5	64.7
St. Kitts/Nevis	2.4	22.7	74.8
St. Lucia	4.1	18.4	77.5
St. Vincent/ Grenadines	7.6	25.3	67
Suriname	10.8	24.4	64.8 (2005 estimate)
Trinidad and Tobago	0.4	58.9	40.7

Source: CIA—The World Factbook 2012 (2011 estimates unless otherwise noted).

growth of tourism has not led to a corresponding growth in agriculture, as was expected (Lewis, 1958; Momsen, 1972; Timms, 2006; Rhiney, 2009). Today agriculture in the region is in decline, with significant implications for food security. Traditional exports like sugar and bananas have not been able to withstand the loss of preferential access to markets in Europe. Sugar in particular has declined dramatically in terms of its importance to foreign exchange earnings, although it still employs a significant portion of the labor force in former slave economies in the region. For example, in St. Kitts and Nevis, sugar accounted for 78 percent of export earnings in 1978, but in 2004 it accounted for just 2 percent of export earnings though it employed 8 percent of the labor force and occupied 30 percent of arable lands (Potter et al., 2004). Domestic food production has declined sharply over the past 15 years in particular, and today the Caribbean is a net importer of food—"a paradise that cannot feed itself" (Ahmed and Afroz, 1996, p. 4). Only Guyana and Belize are net food exporters (Deep Ford and Rawlins, 2007; Caribbean Food and Nutrition Institute, 2007). There is no real consensus as to when the decline in agriculture may have started in the Caribbean, but Ahmed and Afroz (1996) postulate that the turning point for Jamaica may have been after independence in 1962 and that the decline would have intensified during the 1980s under structural adjustment policies. In 1981, according to Ahmed and Afroz, there was a 40 percent increase in food imports due to the government's decision to remove existing import restrictions on a range of foodstuffs that could be produced locally. This resulted in competition from cheaper imports, which in turn led to a 12 percent reduction in domestic agricultural production the following year. Weis (2004) described this decline of domestic food production in Jamaica as a crisis. This is supported by Ahmed and Afroz (1996, p. 7), who submitted that "[n]ot only has Jamaica been unable to expand its agricultural items, but it faces a major food crisis because of declining production of staple food for the local market." In addition, agriculture as a percentage share of GDP and total labor force has also declined in recent times. According to a study done by the World Bank (1993), agriculture's percentage of GDP in Jamaica decreased from 8.3 percent to 5.7 percent between the 1970s and 1985. At the same time, the percentage of the labor force employed in agriculture dropped from 30 to 24 percent.

This chapter explores the role and place of agriculture in Caribbean economies in the context of continuity and change and the implications for regional food security.

1.2 Contextualizing Agriculture in the Caribbean

Fifty years after the English-speaking Caribbean started to rid itself of colonial rule with the independence of Jamaica, and 174 years after emancipation, agriculture in the Caribbean is still fundamentally shaped by the legacy of the plantation system and the colonial political economy in which it originated and flourished (Potter et al., 2004; Timms, 2008). This political economy connected Caribbean economy and society to extraregional markets in the metropolitan countries of Europe, a link that has persisted in the contemporary Caribbean. Plantation crops such as sugarcane throughout the region and other crops like nutmeg, arrowroot, and pimento in, respectively, Grenada, St. Vincent and the Grenadines, and Jamaica were exported, while inputs to agriculture, most of the foods eaten by the masses such as flour and salted fish, and consumer goods were imported. In the plantation system and throughout the colonial era, sugar was king and plantations cultivating sugarcane and other export crops were established on the best arable lands. In the case of sugarcane, these were coastal plains and interior lowlands and alluvial plains. This left only marginal lands for small-scale domestic agriculture. During slavery, the provision grounds made available to slaves to grow some food for themselves were established on the most marginal plantation land that was basically unsuitable for sugarcane cultivation. In 1838, slaves were emancipated throughout the British Empire. Former slaves could continue to work on the plantation for wages; however, desperate to distance themselves from their former lives, many of them chose to leave the sugar estates and established independent communities. Farming and other informal activities became the basis of rural livelihood in these free villages. With all the best lands occupied by sugar plantations, marginal land on hilly terrain and steep slopes was utilized by small-scale farmers. Small-scale domestic agriculture in the Caribbean came to be associated with hillsides and is sometimes described as hillside farming. As part of the dichotomized and dual structure of Caribbean agriculture, sugarcane still occupies the best lands. Sugarcane cultivation was in decline throughout the region even before preferential markets in Europe were dismantled, after which it became even more unprofitable and uncompetitive; sugar is no longer king. In many islands, land that has been taken out of sugarcane cultivation has not been made available to smallholder farmers. In Jamaica, some former sugarcane lands have been put into banana cultivation—another export crop, but which also has a huge domestic market given its critical role as a staple food. Elsewhere, like St. Kitts and Nevis, former sugar plantations have been used for housing developments.

Agriculture in the Caribbean did not originate with European colonization. The first peoples in the region are thought to have been Amerindians, mainly Tainos, Arawaks, and Caribs. They were all agriculturalists to varying degrees and their influences can still be seen in contemporary Caribbean agriculture. Crops like beans, corn, cassava, and pumpkin, which were grown by the first peoples, are still grown around the region today. Other cultural influences can also be identified, including African, East Indian, and European. African influences are found in many of roots and tubers grown across the region. In Guyana and Trinidad, which both have large populations of people of East Indian ancestry, rice is a popular crop. Vegetables such as carrots, onions, and cabbage were introduced from Europe, and breadfruit, also popular throughout the region, was introduced from the Pacific Islands.

1.3 The Role of Agriculture in a Global Context

According to Meijerink and Roza (2007), there appears to be somewhat of a paradox in the role of agriculture in economic development. This is manifested in the fact that while agriculture's contribution to GDP has been steadily declining, the production of some critical crops like cereals has been increasing. It would seem that agriculture's role in the total economy might actually decline as it becomes more successful (Meijerink and Roza, 2007). They suggest that it might be tempting to say that an emphasis on other sectors of the economy at the expense of agriculture facilitates economic growth. But Haggblade (2005) posits that the agricultural sector is still important in economic development, but that the overall economic growth reduces the role of agriculture when measured in terms of contribution to GDP. This analysis makes it clear that assessing the role and contribution of agriculture in isolation from the state of other economic sectors constitutes an incomplete approach. The majority of the world's poor live in rural areas and for these people agriculture is the main source of employment and income. It was estimated that in developing countries, 53 percent of the total workforce was employed in agriculture, with the figure in sub-Saharan Africa being 60 percent (World Development Indicators, 2006).

1.4 The Role of Agriculture in the Caribbean

Agriculture makes several important contributions to Caribbean economies. First, it still contributes significantly to GDP in most countries of the region. In the very small islands its role is not as important. For example, in the Bahamas only approximately 2 percent of the total land area is cultivated, and farming is more accurately defined as gardening and serves mostly subsistence functions. Still, agriculture accounts for 3 percent of GDP. Increased investments in other sectors and continued bias against agriculture, especially the small-scale farming sector, have resulted in increased contributions of other sectors to GDP and a decline in agriculture's contribution. In Antigua/Barbuda, for example, agriculture accounted for 20 percent of GDP in 1961, but with the collapse of the sugar industry and the development of a thriving tourism industry, this declined, and in 2000 agriculture contributed only 4 percent to GDP (Potter et al., 2004). In other countries, export agriculture is an important economic activity and contributes significantly to foreign exchange earnings. This applies to traditional export crops like nutmeg in Grenada, arrowroot in St. Vincent and the Grenadines, rice in Guyana and Trinidad and Tobago, bananas in the Eastern Caribbean and Jamaica, sugar in many islands, and coffee in Jamaica. In some islands like Jamaica, there is a growing category of relatively recent agricultural export crops, mainly roots, tubers, and winter vegetables developed since independence. These have contributed significant foreign exchange earnings in the last two decades and have the potential to be even more important. Agriculture also plays a critical role in food security throughout the region, providing domestic supply of food and energy for the workforce. In this respect it serves to save foreign exchange that would otherwise be spent on food imports.

In most Caribbean countries agriculture is the major economic land-use activity, although it is contributing less and less to GDP. Official data typically indicate single-digit contributions of agriculture to GDP. For example, it is estimated that agriculture contributes 8 percent to Jamaica's GDP, but the Inter-American Institute for Co-operation in Agriculture (IICA) suggests that when backward and forward linkages in the commodity supply chain are accounted for, the contribution of agriculture to GDP is three to seven times higher than that reported in national statistics. IICA studies in Latin America have shown that agriculture's contribution to GDP is usually underestimated. For example, in Argentina, official statistics indicate a 4.6 percent contribution of agriculture to GDP, but this figure increases to 32 percent when forward and backward linkages are considered. In Brazil, the official figure of 4.3 percent rose to 26.2 percent; in Chile, Mexico, and Costa Rica, respectively, official figures of 5 percent, 4.3 percent, and 11.3 percent increased to 32.1 percent, 24.5 percent, and 32.5 percent (IICA, n.d.).

The importance of agriculture is underestimated because of how its contribution is perceived and measured. In the Caribbean and the developing world in general, agriculture's contribution to GDP tends to be measured only in terms of primary production. The value-added elements of agro-processing and the employment function and other linkages are rarely taken into account. It should be noted that the same is true of developed countries. For example, in the United States, the official estimate of agriculture's contribution to GDP is 0.7 percent, but when backward and forward linkages are considered, this increases to more than 8 percent (Meassick, 2004).

The contribution of agriculture therefore tends to be underestimated. For one thing, its importance in agrobusiness is not always accounted for (Meassick, 2004). Also, agricultural professionals and workers are not counted when the role of agriculture is being assessed. Farmers are not the only category of agriculturalists (Meassick, 2004). There are also agro-processors, food inspectors, extension workers, educators, veterinarians, engineers, and salespeople and marketers. The agricultural sector should be assessed in its broadest sense and not narrowly defined in terms of primary production. According to the IICA, agriculture has a threefold role in national development in the Caribbean: food security, social stability, and environmental protection. It is critical to rural development and prosperity as it absorbs surplus labor and provides income and livelihood. Livelihood assets in the rural context are often defined largely in agricultural terms (Campbell, 2011).

According to Potter et al. (2004), employment in agriculture as a percentage of the total labor force has declined in recent years. Among CARICOM countries, the agricultural labor force ranged from 8 percent in Trinidad and Tobago to 20–30 percent in Belize, Dominica, Grenada, and Jamaica (Potter et al., 2004). In terms of land use, agriculture still occupies significant space: about 45 percent in Jamaica and Barbados, and 30 percent in the Organization of Eastern Caribbean States (OECS) countries. On the other end of the spectrum, only 2 percent of land in the Bahamas is cultivated, largely because of terrain and soil constraints. Like the Bahamas, the Cayman Islands and Montserrat have very small percentages of land under agricultural cultivation. In some countries, such as Guyana and Belize, agricultural land use has increased, while in others, such as Grenada and St. Kitts and Nevis, it has declined (see table 1.2). In the former, expansion in rice cultivation led to the increase. In Belize, where less than 7 percent of land is under agriculture, there has been a significant increase in agricultural land in the last 25 years and a doubling since 1961 (Potter et al., 2004).

It was estimated that in 2009 agriculture accounted for 7 percent of GDP in Jamaica but absorbed a significant 21 percent of the employed labor force (CARICOM Secretariat, 2009). It contributed greatly in provision of fresh produce for agroprocessing and also accounted for more than 20 percent of exports in 2004 (CARICOM Secretariat, 2009). Despite the obviously important role of agriculture in the economy, the country still imports 80 percent of the cereal and cereal products used, as well as 12 percent of its meats and a growing volume of fruits and vegetables. Agriculture can play an even greater role through the expansion of aggregate output and value added.

The role of agriculture in the Caribbean has been shaped to a large extent by the declining fortunes of traditional export crops, especially sugar and bananas. Both have suffered as a result of the loss of preferential access to international markets in Europe. In the Eastern Caribbean, the banana

Table 1.2 Changes in agricultural land in CARICOM, 1961–2009

Countries	1961 (percent)	1981 (percent)	2000 (percent)	2009 (percent)
Antigua/Barbuda	23	25	27	29.5
Bahamas	1	1	1	1.4
Barbados	44	44	44	44.2
Belize	3.5	4	6	6.7
Dominica	23	25	23	32.0
Grenada	65	47	35	35.3
Guyana	7	9	9	8.5
Haiti	46	51	51	66.8
Jamaica	49	46	46	41.5
St. Kitts/Nevis	56	45	28	19.2
St. Lucia	28	33	31	18.0
St. Vincent/ Grenadines	26	31	33	25.6
Suriname			0.6	0.5
Trinidad and Tobago	20	25	26	10.5

Source: Adapted from Potter et al. (2004) and World Bank (2010) (data.worldbank.0rg>indicators).

industry has likewise been adversely impacted by loss of preferential markets and competition from cheap banana produced in Central America (Grossman, 1998). Today domestic food crops outperform export crops and represent the driving force behind Caribbean agriculture. In Jamaica, for example, yam cultivation is one of the few bright spots in the agricultural sector (Barker and Beckford, 2006) and makes an important contribution to export earnings.

Agriculture in the Caribbean is also negatively impacted by the ubiquitous natural disasters to which the region is susceptible (McGregor, Barker, and Campbell, 2009; Campbell, 2011; Beckford, 2009). The main ones are meteorological hazards such as hurricanes, droughts, floods, and landslides. Hurricanes get most of the headlines and have the most immediate impact, but the scenario is really one of alternating cycles of hurricanes, drought, and floods, often experienced in the same year. The impact of natural disasters is discussed in chapters 11 and 12 in this volume, but billions of dollars in losses have been incurred by the agricultural sector over the last few decades (McGregor et al., 2009; Campbell, 2011). The last two decades have seen an increase in the frequency and intensity of natural disasters, and the agricultural sector throughout the region has taken a pounding. The time to recover from extreme events is getting shorter and shorter, and this affects stability and consistency in performance of the agricultural sector (Campbell, 2011).

The decline in agriculture in the Caribbean since the middle of the twentieth century is partly a result of the forces of globalization, which have wreaked havoc on the traditional export sector, especially sugar and banana. Potter et al. (2004) suggest that structural change has led to the evolution of service-dominated economies, led by tourism. A major deficiency in agriculture in the Caribbean is the lack of linkages with other sectors. Nowhere is this more evident than in the relationship between agriculture and tourism (Timms, 2006; Rhiney, 2008; Momsen, 1972). As we alluded to earlier the tourism industry in most Caribbean countries is expanding while local agriculture is on the decline, and the region has become a net importer of food (Timms, 2006; Rhiney, 2008; Deep Ford and Rawlins, 2007; Beckford, 2012). The weak linkage between agriculture and tourism in the Caribbean is a long-standing problem (Momsen, 1972). In the 1960s, Nobel Prize laureate Sir Arthur Lewis had proposed an economic theory to forge inter-sectoral linkages, but Lewis's theory was not implemented as the region preferred a model of industrialization by invitation, which focused on foreign external investments to the detriment of local agriculture (Timms, 2006, 2008). Industrial growth and agricultural activities thus took place in isolation from each other (Timms, 2006). An analysis of the relationship between local agriculture and tourism is the subject of Chapter 16.

1.5 Summary

Agriculture still provides the foundation for the economies of the Caribbean. It contributes to many aspects of people's lives, such as food security, employment, and income. Its fundamental role in rural development is well documented, but its importance in urban life should not be underestimated. Agriculture is critical to political and social stability in the Caribbean. This is a role that is not well researched and not well understood. Hunger and food insecurity, often rampant in other parts of the developing world, have been largely unknown in the region, although Haiti continues to lurch from one crisis to another, with implications for food security. Agriculture used to be the backbone of the economy of the Caribbean. With its declining contribution to GDP and the decline in the proportion of people involved in agriculture (narrowly defined as crop production and livestock rearing), some could argue that this is no longer the case. Today manufacturing is becoming increasingly important to many regional economies, tourism and the hospitality sector more generally is currently the main source of income in many countries, and remittances now play a crucial role in many countries' economies. Despite this, we would argue that agriculture is still an essential component of Caribbean economies. It provides food and energy as well as income and employment to many people. Agriculture is more than just primary production, and outside of direct farming there are many people engaged in agriculture through agroprocessing, marketing and distribution, agricultural extension work, transportation, and other areas. Agriculture is important in poverty reduction through food provision and employment and is still a major foreign exchange earner in most islands. Agriculture is also important in maintaining biodiversity in the region, but this role is also under-researched and often misunderstood.

We argue that the development of the Caribbean requires a strong agricultural sector. The status of food security requires a healthy domestic food production sector. Today food security in the region is heavily dependent on international trade (Deep Ford, 2006; Deep Ford and Rawlins, 2007). The current dependence on imported food by the countries in the Caribbean is unsustainable. The small-scale farming sector in many countries plays a critical role in food security and has the potential to make an even bigger impact. This should be a key consideration in agricultural planning and policy in the region.

CHAPTER 2

The Small-Scale Food Farming Sector in the Caribbean: Food Production and the Caribbean Peasantry

2.1 Overview

The responsibility of producing food crops in the Caribbean falls mainly on the shoulders of thousands of small-scale farmers who cultivate small holdings on mostly marginal lands in the interior hilly areas, certain river valleys and flood plains, and the dry southern coastal plains. Most farmers still use the same cultivation methods that were used hundreds of years ago. Most food products are sold, processed, resold, and consumed locally, thus providing the foundation for people's nutrition, incomes, and livelihoods and contributing to rural and national development (Beckford and Bailey, 2009). The small-scale food producers have been able to grow food crops despite enormous challenges facing them(these challenges have been documented by McGregor et al., 2009; Campbell and Beckford, 2009; Beckford and Bailey, 2009; Beckford, 2009; Barker and Beckford, 2008; Beckford et al., 2007). The significance of agriculture in the Caribbean is historical and extends beyond just satisfying household needs. Agriculture's contribution is indispensable to national, community, and household food security. In the 1970s, agricultural policy shifted its focus on food self-sufficiency, as a result of which domestic food production and consumption of locally grown foods were prioritized. Thus, agriculture—which had historically been the backbone of the economy—and the small-scale domestic food sector—which had always been the driver of food security—became even more important.

Emerging from the peasantry in the nineteenth century, the small-scale farming sector developed to become the backbone of the agricultural sector. Over the past two and a half decades, as export of traditional crops like sugarcane and banana has declined, the small-scale farming sector has become the one bright spot on the Caribbean agrarian landscape. With concerns about regional food security increasing over the past one and a half decades, focus has shifted to this sector and efforts have been made to improve food self-sufficiency and the overall food security.

Small-scale producers cultivate farms with an area of 2–4 acres on average, rarely with all their holdings under cultivation at any given time. Throughout the islands, these small-scale cultivators occupy marginal lands in the hilly interior or coastal lowlands as export crops dominate the most arable lands (Barker, 1993). This pattern of land use is a legacy of the slave plantation era, when sugar plantations occupied the best lands in the alluvial plains and coastal lowlands. Even with the decline in sugar exports, these fertile lands in the alluvial valleys and plains have not become available to small-scale food producers.

Small-scale farming in the Caribbean serves both commercial and subsistence purposes. In fact, for many small-scale farmers, agriculture is not just their main source of income; it may be their only source of income. They cultivate a wide range of domestic food crops that are staples in the local diets and cuisine. These include root crops and tubers such as yams, sweet potatoes, and dasheen cassava; vegetables such as cabbage, lettuce, carrots, cucumbers, and tomatoes; condiments such as peppers, onions, and thyme; a wide range of legumes; and a variety of fruits. In Trinidad and Tobago and Guyana, rice is a major crop cultivated by small-scale farmers. Some of this produce is consumed in the home, but significant quantities are sold in local fresh produce markets. Over the past 40 years, Caribbean governments have secured overseas markets for nontraditional export food crops including those mentioned previously in text; this implies that the small-scale farming sector is involved in export trade as well (Potter et al., 2004).

Farming systems in the Caribbean, like tropical small-scale farming systems generally, tend to be quite diverse with locational and situational uniqueness observed between different farming communities in any given island. This is due, in part, to the diverse agro-ecological characteristics and historical and cultural influences. At the same time, certain ubiquities exist in the tropical small-scale farming systems in the Caribbean.

Some of the major characteristics that seem common across different farming systems across the region include aging farming populations with a general absence of younger people, dominance of men in the actual cultivation of crops, dominance of women in the marketing and distribution of domestic food crops, a variety of tenure arrangements, use of traditional farming methods, prevalence of mixed cropping and multiple cropping systems, high rates of farm fragmentation, low level of education among farmers, and an aversion to risks in decision-making (Collymore, 1986; Semple, 1996; Brierley, 1991; Hills and Iton, 1988; Grossman, 1998; Barker et al., 1983). Other ubiquities include the importance of food forests and kitchen gardens and agroforestry techniques in general and rain-fed agriculture. Small-scale farming systems are also characterized by low resource base, lack of access to affordable funding for farm expansion and innovation, lack of research and development, inadequate extension services, and susceptibility to natural hazards.

This chapter will discuss some of these characteristics and analyze their impact on food production, regional food self-sufficiency, and food security.

2.2 Introduction: Historical Context

Understanding the historical context of domestic agriculture in the Caribbean is fundamental to understanding the current state of the small-scale farming sector in the region, the current reliance on imported food and the huge and burgeoning food import bill, and the implications for food security. An important aspect of this is an understanding of the development discourse about the place of domestic agriculture in the development of Caribbean economies. Domestic agriculture in the Caribbean has had a checkered history in terms of its perceived role and importance in development. It can be argued that both development policy and theory in the region have tended to favor export agriculture and industrial development to the detriment of domestic agriculture. Domestic agriculture is perceived as an unviable development option, and export agriculture with its supposed comparative advantage is a more sustainable economic driver (Timms, 2008; Figueroa, 1996). Timms posits that efforts to position domestic food production as an important development imperative has been repeatedly hindered by internal and external forces (2008).

Timms argues that "[t]he application of development theory in the Caribbean has legitimized industrialization and export agriculture at the expense of agricultural production for the domestic market, maintaining the colonial legacy of a plantation economy" (2008, p. 101). The neglect of domestic agriculture has contributed to increasing dependence on imported food with serious implications for regional food security. This reflects the legacy of the plantation system and colonialism (Timms, 2008; LeFranc, 2006). The political economy of the British slave economies in the Caribbean was driven by the principle of mercantilism in which the island colonies were producers and exporters of primary products and raw materials for the British market and importers of British manufactured goods (Augier et al., 1961; Greenwood and Hamber, 1980; Williams, 1970). During slavery, slaves were allowed to grow some food on land provided by their owners, but this was typically marginal land unsuitable for sugar cultivation. Sustainable food production in the colonies was not prioritized as it would detract from the primary purpose of the colonies, which was to supply the metropole with tropical products, especially sugar (Williams, 1970). Dependence on imported food and an emphasis on export-oriented production were therefore created by the economic and political order of the plantation era. Renowned Caribbean statesman, writer, and former prime minister of Trinidad and Tobago Sir Eric Williams indicated that calls for diversifying the economy of the Caribbean came as early as the end of the seventeenth century, but the British parliament rejected a proposal to ban food imports in 1698 in an effort to stimulate food production in the British Caribbean colonies (Williams, 1970). This continued after emancipation in 1838 with restricted access to land for ex-slaves, forcing large numbers of them to continue working for the plantation in the furtherance of the mercantile interests of the colonial master (Mandle, 1982).

In the nineteenth century, with the Industrial Revolution in Britain, free trade economics replaced mercantilism, but this had no significant impact on food production in the Caribbean (Greenwood and Hamber, 1980). The focus was still on export production of sugar and importation of food. It was not until the end of the nineteenth century when competition from other sugar producers, including European beet sugar, led to the decline of the British Caribbean sugar industry that the diversification of the Caribbean economy became an imperative of the imperial government (Greenwood and Hamber, 1980). The impetus for change came out of the British Royal Commissions of 1882 and 1896. The 1882 Commission found that with the exception of St. Lucia, where exports of tropical products, especially sugar, had trebled in the previous 30 years, the export agriculture economy was backward and outdated (Augier et al., 1961; Greenwood and Hamber, 1980; Williams, 1970). The 1882 Commission recommended among other things the diversification of the economy and specifically the encouragement of small-scale farming (Greenwood and Hamber, 1980). The 1896 Norman Commission was even stronger in its repudiation of the economic structure of the British Caribbean colonies. The commission was a response to plummeting sugar prices and the political philosophy of the British Secretary of State Joseph Chamberlain, who suggested that the development of the colonies should reflect the interests of both the colonies and Great Britain (Greenwood and Hamber, 1980). The Norman Commission blasted the sugar industry recommending that with the exception of Antigua/Barbuda, Barbados, and St. Kitts/Nevis, sugar should not remain a staple crop and diversification should take place. This should have opened up the way for domestic food crops, and although this sector did see an expansion, it has to be said that the alternative crops grown were still largely oriented toward the export market (Greenwood and Hamber, 1980). Alternative crops that became important included cotton (Barbados, St. Kitts/Nevis, Antigua/Barbuda, and Montserrat); spices such as nutmeg in Grenada, ginger and pimento in Jamaica, and arrowroot in St. Vincent and the Grenadines; cocoa in Jamaica and Trinidad and Tobago; rice in Trinidad and Tobago and Guyana; coconut plantations in Jamaica and Trinidad and Tobago; lime in St. Lucia and Dominica; and banana in the Windward Islands and Jamaica (Greenwood and Hamber, 1980). Although these crops were mainly for export, they were grown significantly by small-scale farmers and so presented the first challenge to the hegemonic dominance of the plantation system.

Timms (2008) suggested that the recommendations of the Norman Commission were not implemented and domestic agriculture received very little attention. After emancipation in 1838, an independent peasantry developed, which provided impetus to domestic food production. Timms (2008) suggests that even then domestic agriculture was basically characterized by subsistence practices in modes of production and marketing. Still it is generally recognized that the small-scale farming sector, which became the backbone of the Caribbean economies, emerged first from the communities of free persons, which sprung up after 1838. Witter and Beckford (1980) (cited by Timms, 2008) claim that it was the establishment of a local network for the distribution and marketing of domestic food crops that increased the contribution of the sector, which stimulated the development of road networks from the rural interior to the urban commercial centers. But as Timms quite rightly points out, the development of a local domestic agriculture through the local peasantry did not change the economic structure of the Caribbean from exporters of raw materials and primary products and importers of food (2008). Potter et al. (2004) have shown how Caribbean economies have evolved from and have been shaped by the plantation system. This view is shared by LeFranc (2006), who posited that small-scale farming in the Caribbean evolved from the plantation economy and was actually a reaction to it and that this can be used to explain the behavior of smallholder farmers in the region.

2.3 Domestic Agriculture and Development in the Caribbean

After World War II, Caribbean economists and development thinkers began to consider and write about context and situational directions of Caribbean development. This was led by Nobel laureate Sir Arthur Lewis, who articulated a Caribbean economy with a dual structure comprising a large subsistence sector and a smaller capitalist sector, with symbiotic linkages between them (Lewis, 1954). The subsistence sector would provide cheap labor to the capitalist sector, which would reciprocate by providing markets for the products from the subsistence sector. In Lewis's model of development, export-oriented industrialization would be the focus, with a supporting and complementary role played by domestic agriculture. His *industrialization by invitation* model would use the availability of cheap labor from the subsistence sector to attract foreign investments and industries. As agriculture modernized, the subsistence sector would contract and become more productive.

Lewis's development model suggested that development should start outside of agriculture. He theorized that growth in domestic agriculture without other sectors being strong would be counter-productive, and there would be no markets for produce. He also felt that agriculture could not be the basis for development because of the difficulties in getting favorable agricultural terms of trade (Lewis, 1958). The development of a non-agricultural sector would induce demand for food and create intersectoral linkages. Lewis's balanced approach to development was not adhered to in the Caribbean (Timms, 2008). Industrialization was pursued at the expense of domestic agriculture; intersectoral linkages did not materialize as Lewis had envisaged and local economies were marked by weak markets and dependence on foreign investments, foreign markets, and foreign food (Demas, 1975; Axline, 1984).

Best (1976) wrote that the industrialization by invitation experiment did not work in the Caribbean. Criticisms of the model were largely based on the fact that it purportedly led to regional dependence on external forces when the regional agenda should have been focused on increasing economic and political independence (Best 1976; Girvan and Jefferson, 1971; Beckford, 1972). Critics called for attention to domestic food production stating the obvious that without local food production, foreign exchange earnings would go toward paying for imported food (Beckford, 1972). It was argued that the domestic food sector was important to development and that small farmers were more productive than those who worked on plantations (Beckford, 1975; Marshall, 1985). They should therefore be supported by providing access to good quality land, technical advice, access to financial resources, subsidies, and protection from foreign competition (Bernal et al., 1984).

In the past 50 years, there has been a greater recognition of the role of domestic food production in Caribbean development. However, despite this recognition the domestic food production sector is still marginalized. This has resulted in domestic agriculture being neglected and a growing dependence on imported food. The global food crisis of 2008 brought the issue of food security in the region to the attention of a complacent society and led to a flurry of activities and initiatives to address growing food insecurity in the region through increased local production and reduced dependence on imported food.

2.4 Characteristics of Smallholder Farming Systems in the Caribbean

Understanding the internal characteristics and dynamics of local farming systems is important for addressing the issue of food security and nutrition (Campbell, 2011). In this section we briefly discuss the farming systems and the socio-demographic and agronomic characteristics of smallholder farming in the Caribbean. Socio-demographic characteristics include factors such as age, education, gender, farming status, and number of children as well as household characteristics (including its size, ownership, types of productive activities, and off-farm income sources). Agronomic characteristics include farm size and ownership, number of farm plots, cropping systems (past and present), and farming practices.

2.4.1 Structure of Small-Scale Farming

Agriculture in the Caribbean region is characterized by a deep structural dualism (Barker, 1993). There is a small-scale domestic food production sector alongside an export-oriented sector. The former concentrates on staple food crops including a wide range of roots and tubers (yams, cassava, potatoes, dasheen, coco yam), vegetables (carrots, cabbage, callaloo, tomatoes, pumpkin, cucumber), fruits (melons, bananas, plantains), legumes, and condiments (onion, peppers, skellion, thyme, etc.). These are grown mainly for the local market but some have become important export crops (Beckford and Bailey, 2009). The export sector is dominated by crops such as sugarcane, banana, cocoa, coffee, citrus fruits, nutmegs, and pimento. This dualism has historical roots and is a legacy of the colonial era. Before emancipation there was no domestic food production sector although slaves did grow some food on their provision grounds and sold or traded the surplus. The economy was geared toward production of crops for export and food was imported. Emancipation led to the development of a rural peasantry and the growth of a significant domestic food sector over time. The domestic food sector existed in isolation from the entrenched plantation sector and in competition with it for resources. The export sector has been the big winner in this competition. The domestic sector suffers from difficulties in obtaining financial credit, lack of research, poor marketing and distribution infrastructure, and lack of extension services.

2.4.2 Modes of Production

The small-scale farming sector in the Caribbean has always been largely responsible for most of the production of domestic food, which forms the foundation of nutrition and daily diets and is thus fundamental to food security and economic development. It is also responsible for a significant percentage of export crop production across the Caribbean. Throughout the region small-scale farmers are actively involved in the production of export crops such as banana in the Windward Islands and Jamaica, sugarcane throughout the region, spices like nutmeg in Grenada and pimento in Jamaica, coffee in Jamaica, and citrus fruits. There is a tendency to think of Caribbean small-scale farmers as subsistence cultivators, a factor that stems largely from the essentialization of smallholder farmers in the developing world in the academic literature. In other parts of the developing world, particularly tropical Africa and subtropical Asia, smallholder farming is mainly a subsistence activity, but this is not the case in the Caribbean. LeFranc (2006) points out that it is difficult to find places in the Caribbean where small-scale farming is primarily a subsistence activity. Potter et al. (2004) support this view, writing that few smallholder farmers in the region produce solely for subsistence. Since the earliest days of food cultivation in the colonial era there has been a long tradition of commercial farming by small-scale food producers in the Caribbean. For example, during slavery surplus food grown by slaves on their provision grounds was sold or otherwise traded in local markets (Mintz, 1985). The contribution of smallholder farmers to export agriculture in the region is longstanding as well. According to Satchell (1990), during the construction of the Panama Canal between 1882 and 1904, in which thousands of West Indians were employed, root crops were exported from the Caribbean to feed the laborers.

In contemporary Caribbean economies, small-scale farmers are major players in export agriculture with crops like bananas in the Windward Islands and Jamaica; roots and tubers, vegetables, and fruits in Jamaica; and rice in Trinidad and Tobago and Guyana. Many of the crops grown in the region are important staple foods and are also components of a robust internal marketing and distribution network in all the CARICOM countries and are also important in inter-island trade in the eastern Caribbean. One must therefore be careful in ascribing the title *peasant* to Caribbean small-scale farmers even though production modes are still characterized by low technological, labor-intensive, and traditional practices characteristic of peasant economies in Africa, Asia, and Latin America. Mintz (1974) used the term "reconstituted" to describe Caribbean rural farming populations to reflect the influences of urbanization, commercial farming, exposure to and involvement in international agricultural trade, and other social and economic relations not traditionally associated with conventional peasantries. One fundamental feature of all peasantries that also characterizes small-scale farming sector in the Caribbean is its dependence on the centers of economic and political power. These relationships are asymmetrical in terms of the dynamics of power, which govern them (LeFranc, 2006).

Small-scale farming in the Caribbean is resource poor. Farmers cultivate small plots of land using traditional cultivation techniques and with little access to financial resources and with poor infrastructure. Farmers usually do not have access to much financial resources from formal financial institutions like banks and credit unions. The main sources of financial resources include personal funds, income from off-farm employment, and remittances. Some small-scale farmers do manage to obtain capital from formal institutions, but most do not have the kind of collateral often required by financial institutions and won't consider risking houses and land.

Small-scale agriculture and domestic food cultivation in the region generally occur under rain-fed conditions. Most small-scale farmers lack the resources to afford irrigation schemes. Smallholder farming is therefore highly susceptible to the vagaries of weather and climate, and the onset of drought have had severe impact on the domestic food production sector. Some small-scale farmers have put in homemade irrigation systems including drip irrigation (Campbell, 2011). In Jamaica some farmers have set up sprinkler systems connected to domestic water supply (Campbell, 2011) and utilize canals belonging to sugar plantations. But most farmers must wait for rain or use rudimentary watering systems that are usually manual. For example, research in Jamaica has documented the use of watering techniques by some local food farmers in which water storage containers are scattered strategically on farm plots and watering cans or simple small containers are used to collect water and apply to individual plants (Campbell, 2011) (see figures 2.1 and 2.2).

2.4.3 Socio-demographic Characteristics

The small-scale farming sector in the Caribbean can be aptly described as aging. For example, the average age of Jamaican farmers has consistently remained in the 50–60 age group for the past half a century (Campbell, 2011). In a recent study it was found that the average farmer in southern St. Elizabeth Parish was 52 years old.

Farmers' *education level* is often linked to important elements of household livelihood security. It is generally assumed that farmers who have higher levels of education are better equipped to access resources and possess a

Figure 2.1 Water storage containers in dry land farming

Figure 2.2 Hand watering of individual plants

greater capacity to utilize them. However, the reality is that the average Caribbean domestic food farmer has a low level of education with many having attained only primary-level education. In Campbell's research referenced earlier in text, 45 percent of the sample identified primary/all-age school as their highest level of education, 23 percent stated secondary/high school, and 13 percent received some form of tertiary-level education. Importantly, 19 percent of the sample indicated a lack of formal education.

Linked to the issue of education is the *source of farming knowledge.* Given the low level of educational attainment, it is not surprising that informal education is the main source of farmers' knowledge. The main sources include parents and other farmers, trial and error through farm-level experimentation and adaptation, and extension officers (Campbell, 2011). Four percent of farmers identified formal education as their main source of farming knowledge in Campbell's study. For a significant number of farmers, farming is a default occupation that they entered into because of lack of real prospects in other occupations that require higher educational requirements. While most farmers lack formal education and training, many of them have managed to acquire various nonfarm skills via informal apprenticeship. This has enabled them to create alternative sources of income. Off-farm employment is a feature of smallholder farming in the Caribbean. Many farmers operate on a part-time basis for a number of important reasons. Most farmers engage in off-farm work to supplement farm income. Supplemental employment is important in providing working capital for smallholders. Limited farm size, high cost of farm input, and other resource constraints also determined part-time status in farming.

Both men and women are involved in food production in the Caribbean. Men are dominant in actual cultivation of food, but women are also cultivators and dominate in the marketing and distribution of food. In many households an adult male farmer operates on a full-time basis and his wife on a part-time basis, cultivating a small garden near the house while having a job off-farm. This serves to re-emphasize the important role of female farmers in ensuring stability and livelihood security at the household level, a point that will be further developed in Chapter 7.

Farm Characteristics
Activities at the farm level are often shaped by demographic and household characteristics. The basic idea is that, while households and farms are separate units of analyses and for the most part are studied as such, it is important to recognize that changes in the former can significantly alter the latter and vice versa (Campbell, 2011). As such, there has been a thrust toward creating integrated and systematic approaches for agricultural research that takes

into account both farm-level and household characteristics. Important farm-level characteristics include land tenure, farm size (and number of plots), and cropping systems.

2.4.4 Land-Use and Tenure Systems

Contemporary land tenure and land-use patterns in the Caribbean are rooted in the plantation system. For many decades, suitable agricultural lands have been dominated by large-scale production of crops such as sugarcane, coffee, and banana, while small-scale production has been relegated to the infertile fringes of these estates. Small-scale farmers established various forms of land tenure arrangements with estate owners to access such lands. After many decades of government policies geared toward reducing land tenure insecurity among small farmers, access to good-quality and sufficient farmland remains a persistent problem.

Small-scale farming is characterized by multiple tenure arrangements including owning, leasing, renting, and informal user rights relations. Illegal use through squatting is also practiced. Customary tenure is also common in the region. The most common would be family land in Jamaica and common lands in Barbuda. Other examples can be found among the Amerindian shifting cultivators in Belize and Guyana (Potter et al., 2004). These forms of land tenure system emerged during the post-emancipation period as a means of resisting and accommodating the plantation system (Besson, 1995).

There is not always an accurate picture of the proportion of small-scale farmers who operate on family land and other versions of customary tenure. Campbell (2011) found that more than 25 percent of his sample carried out most of their production on family land, which reinforces the important role of customary land tenure systems in reducing land insecurity. Crichlow (1994, p. 83) posits: " . . . without the existence of family land, a large number of poor people, particularly women, would not have had access to land"; thus "the possession of family land saved them from a state of landlessness." Other forms of land tenure arrangements such as sharecropping, rented/leased, and squatting (settled) are also important means through which farmers gain access to land. While most farmers have only one main type of land tenure arrangement, farmers sometimes engage in multiple tenure arrangements to increase access to land in varying ecological niches.

2.4.5 Utilization of Land

Despite complaints about lack of land as a major constraint to production, small-scale farming is characterized by under-utilization of available land.

At any given time, a considerable amount of available land is uncultivated. Many small-scale, resource-poor farmers cannot afford to cultivate all the land at their disposal. Many farmers also respond to market and environmental factors and deliberately adjust cultivation at various times. The cycle of scaling up and scaling down production subsequent to good and poor seasons, respectively, is a clear indication of the uncertainties that characterize crop production in this region.

2.4.6 Farm Size and Fragmentation

Smallholder farming systems in the Caribbean are often characterized by high land fragmentation indices. Farmers typically cultivate more than one plot of land, and research shows that smaller farmers have higher rates of farm fragmentation. Campbell (2011) found a significant and moderate relationship (0.362) between farm size and the number of farm plots. This implies that larger farm sizes tend to correspond more frequently with a higher number of farm plots and, likewise, smaller farm sizes with a lower number of farm plots. Hence, the idea of large farms having fewer plots than small ones does not hold true for this area. Sixty-one percent of farmers operated on two acres or less with 45 percent doing so on three or more farm plots. Igbozurike (1976) also found a relationship between farm size and fragmentation with smaller farmers having more cultivated plots. The level of fragmentation is affected because farmers often have multiple tenure arrangements to satisfy their demand for land, and this often results in the cultivation of multiple plots. It should be noted that fragmentation is also used as a deliberate strategy to make use of unique ecological niches. We discuss this further in Chapter 13.

2.4.7 Cropping Systems

Cropping systems in the Caribbean are dynamic and responsive to personal, economic, and environmental factors. Farmers typically grow one or two main crops, in addition to a variety of other crops. This is often a means of spreading risks in a resource-poor environment. Mixed cropping is a feature of small-scale farming systems in the Caribbean. Farm plots can have as many as six different crops growing together. These often look like a haphazard unsystematic arrangement of crops thrown together but there is a method to this seeming madness. Through traditional knowledge and farm-level experimentation, farmers have developed a keen sense of the symbiotic relationship between different combinations of crops. Some plants are complementary to each other, while others are not. Our discussion in Chapter 12 of this volume explores this aspect of small-scale farming systems in the Caribbean.

Farming communities and different regions often become known for specializing in certain crops. Most areas are conducive to the growth of a variety of crops, but most have some main crops that are widely cultivated and found on most farms in an area. For example, south St. Elizabeth is known for vegetable and fruits and south Trelawny for yams in Jamaica, while in St. Vincent the Mesopotamia Valley is a famous banana-growing area. Farmers tend to cultivate the same crops throughout their lifetime if they remain in the same area. Sometimes new crops will be tried to take advantage of market opportunities, but Caribbean farmers are not known for their responsiveness to new unproven ideas. Caribbean farmers are often described as risk minimizers with a general aversion to risk in the context of uncertainty (Beckford, 2002; Spence, 1996). Farmers display patterns that may be described as *risk appraisal* and *adaptation appraisal* (Grothmann and Patt, 2005, p. 203, cited in Campbell, 2011). They describe risk appraisal as situations where a "person assesses a threat's probability and damage potential to things he or she values, under the condition of no change in his or her own behavior." Adaptation appraisal, on the other hand, occurs when "a person evaluates his or her ability to avert being harmed by a threat, along with the cost of taking such action." Campbell (2011) explains that farmers often assess the likelihood of production failure from a change in cash crop as well as their ability to recover in such event. The types of crops grown and crop combinations used in the Caribbean reflect other priorities than economics (LeFranc, 2006). For example, some tree crops were traditionally grown for shade and as windbreakers. Fallowing and crop rotation are the two main strategies used by farmers to maintain soil fertility and structure. Cropping systems are generally characterized by the integration of small livestock and poultry rearing with crop cultivation. Livestock mainly include pigs, goats, and sheep.

CHAPTER 3

The State of Food Security in the Caribbean: Issues and Challenges

3.1 Overview

It could be argued that while the issue of food security is a global concern, Caribbean peoples and governments are just now fully grasping its significance for the region. Through the mass media, many people in the Caribbean are aware of the chronic food shortages in other parts of the world and the phenomenon of hunger. With the exception of perhaps Haiti, however, the sensationalized and dramatic incidence of food shortages and hunger seen especially in parts of Asia and sub-Saharan Africa is absent from the Caribbean. This has perhaps led to an underestimation of the problem of food insecurity in the region and a complacent attitude. There is serious doubt, for example, about the extent to which there is accurate knowledge and full understanding of the extent of hunger in the region. We argue that given certain realities, food security in the Caribbean is at best precarious. These realities include impact of climate change and the ubiquitous natural hazards, the decline in local food production in the region; the growing dependence on imported food, the estrangement of Caribbean populations from their traditional foods and diets and a movement toward North American–style fast food diets, and disruptions in the economic base and livelihoods in Caribbean rural communities. Food security in the Caribbean requires the revitalization of sustainable local food production systems.

The region has made some strides in addressing food security concerns, but there is still a great deal of work to be done from the standpoint of policy but also at the grassroots level. Erwin Larocque, CARICOM's Secretariat's assistant secretary-general for Regional Trade and Economic Integration, underscored this in highlighting two pressing regional food security issues. First, he made reference to the impact of international developments on the

Caribbean's ability to be self-sufficient in food production and internationally competitive to afford necessary imports. Second, he stressed the need to address food security issues in the context of the Millennium Development Goals. He stressed that the region's dependence on imported food made it vulnerable and noted the increase in diet-related diseases. In the CARICOM region, only Belize and Guyana are currently net exporters of food, and the gap between food imports and agricultural exports for other territories is over 60 percent—over 80 percent for Antigua and St. Kitts/Nevis (Deep Ford and Rawlins, 2007, cited by Barker et al., 2009).

This chapter analyzes the state of food security in the Caribbean. We examine the problem in the context of the various dimensions of food security by exploring issues such as the decline in domestic food production, the estrangement of Caribbean people from local traditional diets and rising North American taste, increasing rates of diet-related diseases, and increase in poverty, which affect entitlements.

3.2 Conceptualizing of Food Security

This section discusses some key concepts of food security that will frame the discussion of the issue in the Caribbean context. The Food and Agricultural Organization (FAO) defines food security as a condition where "all people at all times, have physical and economic access to sufficient, safe and nutritious food to meet their dietary needs and food preferences for an active and healthy life" (World Food Summit, 1996). Four broad dimensions of food security are usually identified—availability, or the supply of food in an area; access, or the physical and economic ability of people to obtain food; utilization, or the proper consumption of food; and stability, or the sustainability of food supplies (World Food Program, 2009). *Availability* speaks to the supply of food and is influenced by factors such as food production, stockpiled food reserves, and trade (FAO, 2008). Aspects of food availability include the agroclimatic essentials of crop and animal production and the sociocultural and economic milieu in which farmers operate (Schmidhuber and Tubiello, 2007). The second dimension, *access*, addresses the ability of individuals and households to purchase food. It takes into consideration the availability of financial resources to acquire adequate food in terms of both quantity and quality. Concerns about access take cognizance of the fact that availability of adequate food at the national or international level does not guarantee individual or household food security (FAO, 2008; Schmidhuber and Tubiello, 2007). The issue of entitlements is therefore critical (Sen, 1981). Entitlements may be defined as "the set of those commodity bundles over which a person can establish command given the legal political, economic and social arrangements of the community of which he or she is a member" (Schmidhuber and Tubiello, 2008, p. 19703). The dimension of *utilization* is closely related to consumption patterns and behavior, which impact nutritional status and hence health and productivity. It is also related to food safety, preparation, and diversity in diets (FAO, 2008; Schmidhuber and Tubiello, 2007). The fourth dimension, *stability*, refers to long-term consistency in the other three dimensions. It accounts for the reality of individual or households losing access and becoming food insecure periodically, seasonally, temporarily, or permanently (FAO, 2008; Schmidhuber and Tubiello, 2007). Food security objectives cannot be genuinely met unless these four dimensions are concurrently fulfilled (FAO, 2008).

Food insecurity may be described as *chronic, transitory*, or *seasonal* (FAO, 2008). Chronic food insecurity, which occurs when individuals cannot meet minimum food needs, is long term and sustained and results from persistent poverty and lack of assets and resources. In contrast, transitory food insecurity is short term and temporary, occurring suddenly when the ability of people to produce and access food diminishes as a result of factors such as variations in food production and supply, increases in food prices, natural disasters, and reduction in income. Seasonal food insecurity is generally predictable given that it results from a regular and cyclical pattern of inadequate availability and access to food, which may be caused by seasonal climatic variations, cropping patterns, and availability of work (FAO, 2008).

The final term, which will be central to this discussion, is *vulnerability*, defined here as conditions that create susceptibility to food insecurity. In other words, the concept refers to possible and future threats to food security even though food security may prevail at the present time. This concept underscores the fact that food security is not a static condition but is dynamic and is influenced by risk factors, the ability of people to manage risks, and vulnerability to certain events (FAO, 2008).

The world produces enough food to provide all of its more than 6 billion inhabitants with sufficient daily nutrition (FAO, 2008). Despite this, nearly 1 billion people qualify as being hungry, and between 2007 and 2008, some 115 million people were added to the global figures of the chronically hungry (Josette Sheeran, Executive Director, World Food Program, 2009). The world's poorest people continue to face great challenges in food security (Chen and Ravillion, 2004). In this context the global food crisis and the prevalence of hunger is indeed a paradox: the paradox of hunger amidst plenty. Food insecurity is the absence of food security, implying that hunger exists as a result of problems with availability, access, and utilization or that there is susceptibility to hunger in the future (World Food Program, 2009).

3.3 State of Food Security in the Caribbean

Food security and insecurity in the Caribbean are affected by several major factors: (i) declines in productivity of land, labor, and management in the agricultural sector resulting in a weakening capacity to supply food competitively; (ii) decline in earnings from traditional export crops resulting in a reduced ability to purchase food; (iii) the erosion and threatened loss of trade preferences for traditional export crops, the earnings of which are used to buy imported food; (iv) the very high dependence on imported food and the uncertainty of food arrival associated with external shocks; (v) the increasing incidents of pockets of poverty, which affects peoples' access to food; and (vi) concerns over the association of the high use of imported foods and growing incidents of diet-related diseases as people become estranged from local traditional foods and environment and adopt North American foods and lifestyles. These issues are all manifested in Jamaica, where domestic food production has plummeted from the halcyon period of the mid-1990s, when food production peaked over 650,000 tons (Beckford, 2000).

Food production in the Caribbean is mainly the domain of thousands of small-scale farmers who cultivate small holdings on mostly marginal land. Most of this food is sold, processed, resold, and consumed locally, thus providing the foundation of people's nutrition, incomes, and livelihoods and contributing to rural and national development (Beckford and Bailey, 2009). This is achieved despite enormous documented challenges facing small-scale food producers (McGregor et al., 2009; Campbell and Beckford, 2009; Beckford and Bailey, 2009; Beckford, 2009; Barker and Beckford, 2008; Beckford et al., 2007). The significance of agriculture in the region is historical and goes beyond satisfying household needs. It makes an indispensable contribution to national, community, and household food security. Agricultural policy in the 1970s shifted its focus on food self-sufficiency in which domestic food production and consumption of locally grown foods were prioritized. Agriculture, which had historically been the backbone of the economy, and the small-scale domestic food sector, which had always been the driver of food security, became even more important. Farming in the region is still dominated by traditional farming techniques.

Despite the importance of agriculture and the small-scale food production sector in particular, the CARICOM region, except Guyana and Belize, has seen dramatic reduction in food output and has become a net importer of food (Timms, 2006; Deep Ford and Rawlins, 2007; CFNI/FAO, 2007). In his welcoming remarks to a forum on agriculture, food production, and food security in the Caribbean and Pacific regions in 2005, CARICOM secretary-general Edwin Carrington pointed out that up to the mid-1980s, the CARICOM region was a net exporter of food but had since become a net importer of food. Food insecurity in CARICOM (with the exception of Haiti) tends to be understated. It is often construed simply in terms of availability and complacency exists because again with the exception of Haiti, the dramatic and sensationalized incidence of hunger often seen in parts of the developing world is absent (Beckford, 2012). However, it is now acknowledged that the region faces urgent and significant food security challenges. CARICOM has been experiencing declining agricultural productivity, decreasing earnings from traditional export crops, a high and growing dependence on imported food, increasing levels of poverty, and increases in diet-related diseases like diabetes, hypertension, and obesity (Caribbean Food and Nutrition Institute/FAO, 2007).

According to the Caribbean Food and Nutrition Institute, the World Food Summit's definition of food security is especially relevant to the Caribbean, given that countries "are experiencing rapid dietary/nutritional, epidemiological and demographical transitions" (CFNI/FAO, 2007, p. 15). Dietary transition is characterized by a shift from traditional diets based on indigenous staples of local fruits, vegetables, and legumes to diets based on more processed foods with more fats and sugars (CFNI/FAO, 2007, p. 15). From an epidemiology standpoint, "nutrition related non-communicable diseases such as diabetes, high blood pressure, stroke, heart diseases, and cancer have replaced malnutrition and infectious diseases as the major public health problems" (CFNI/FAO, 2007, p. 16). Obesity has been increasing in the region over the past 25 years, and this has precipitated the increases in nutrition-related chronic diseases (CFNI/FAO, 2007). These problems are related to the rise of urbanization in the region (CFNI/FAO, 2007).

Assessed in the context of the various dimensions of food security, the state of food security in the Caribbean becomes clearer. Famine and hunger, which characterize much of sub-Saharan Africa and parts of Asia, are typically not associated with the Caribbean—with the notable exception of Haiti. However, in light of declining food production, over reliance on imported food, growing poverty, and the growing incidence of diet-related diseases, food security in the region may be described as precarious or, to use a technical food security term, *vulnerable* (Beckford and Bailey, 2009). In the case of availability it might be argued that the Caribbean region is safe. Food availability is determined by local production, agroprocessing, food aid, food trade, and food reserves or stockpiles. We have already seen that local production has declined significantly over the past two to three decades and CARICOM countries as a whole have moved from net exporters to net importers of food. The region is now very heavily dependent on food imports to meet its food needs (Beckford and Bailey, 2009). In 2004 the region's food import bill was US$2.409 billion up from US$1.553 billion in 1995 (CFNI/FAO, 2007). In 2006 Jamaica alone imported some US$1.64 billion worth of food, which was half the country's total import bill (Beckford and Bailey, 2009).

The situation is similar in many other CARICOM states and the develop-ing world, where markets have been opened up through trade liberalization (Short, 2000; Spitz, 2002; Walelign, 2002). We argue that this dependence on food imports constitutes a major threat to Caribbean food security. First of all, purely from a livelihood perspective, it does immeasurable damage to local producers and rural development. Faced with unfair competition and the dumping of cheap, heavily subsidized food mainly from the United States, many farm families experience difficulty providing a satisfactory livelihood for themselves (Beckford and Bailey, 2009; CIOEC, 2003; Via Campasina, 1996, 2003; UNDP, 2005). Most of the imported food to the Caribbean comes from the United States, where heavily subsidized production enables farmers to sell for less than the cost of production (Windfuhr, 2002, 2003; Windfuhr and Jonsen, 2005). Local farmers are therefore forced into unfa-vorable, oftentimes insurmountable, competitive situations, and in Jamaica, for example, many have succumbed to this dumping of cheap exports and gone out of business (Beckford and Bailey, 2009).

It might be argued that opening up local markets to international compe-tition is beneficial to consumers through the lower prices that should accrue as a result of competition and that this should stimulate more efficient local production, thereby providing even greater access to affordable food. The problem with this argument is that unfettered competition from heavily sub-sidized foreign food producers has coincided with the removal of subsidies from local producers, creating an uneven playing field (Beckford, 2012). The over-reliance on imported food raises other obvious dangers as well. For one thing, the structure of the world economy means that external shocks often reverberate throughout the system with devastating consequences for the most vulnerable nations and people. Market stability is a major concern here as it causes uncertainty of supplies and rising prices, which could result in food shortages in the region. Given the dependence of the Caribbean on food from the United States, terrorist attacks on the food system in America could have serious implications for Caribbean food security (Beckford, 2012; Beckford and Bailey, 2009). The US$26 Million Food Security Project imple-mented in 2002 was partly in response to the near–food crisis in some Caribbean countries in the aftermath of the September 11 terrorist attacks on the United States. Food safety is an important part of food security, which perhaps does not get enough attention in the literature on the subject. In the context of the Caribbean, we suggest that this should be a real concern. Apart from the threat to safe food posed by terrorist attacks, the long distance traveled by food imported from distant places significantly increases the risk of food being contaminated (Halweil, 2005). Beckford and Bailey (2009) pointed out that the longer the food travels, the more it changes hands, thus increasing the risks of contamination. Halweil (2005) argues that the centralized nature of American food production and processing increases the risks of contaminated food, while the size and uniformity of farm operations create ideal conditions for the rapid spread of diseases. The dependence on foreign food and especially American food is therefore cause for concern.

In terms of the food security dimension of access, the rising and persistent poverty among the population in some Caribbean countries has been a concern that has featured in the regional food security strategy. While it is no doubt true that poverty, hunger, and starvation, common in some parts of the developing world, are largely unknown in the Caribbean, it is also true that many households and individuals in the region experience hunger from time to time, with the rising use of food stamps and other food aid programs being observed. The CFNI/FAO (2007) reports that poverty rates in the region have been declining, but countries such as Haiti and Guyana and a few in the Organization of Eastern Caribbean States continue to have high rates. Income distribution inequity in some countries of the region is among the highest in the world (CFNI/FAO, 2007).

Food security in the region is also greatly impacted by natural disasters. These are a ubiquitous part of life in the Caribbean and exert compelling influence over food and agriculture. Hurricanes, floods, and droughts in particular have combined in cyclical patterns to impact agriculture and food in ways that could be appropriately described as crippling. Food availability is most conspicuously affected, but nutrition suffers as well, and because agriculture is closely tied to income and livelihoods in the region, access to food is always a victim of natural disasters. The fact that two chapters in this volume are dedicated to issues of weather and climate underscores the critical role of climate change in food security in the Caribbean (see chapters 10 and 11).

3.4 Summary

Food security in the Caribbean must be analyzed in the context of the realities of most of the countries in the region, including "small size, vulnerability to natural disasters and a challenging new economic environment characterized by, inter alia, international competitiveness, loss of preferential markets for traditional agriculture" (CFNI/FAO, 2007, p. 1).

In terms of food availability, the region has actually seen an increase in caloric intake, which has been above the recommended rates since the 1960s, characterized by increases in carbohydrates, proteins, and fats (CFNI/FAO, 2007). On the other hand, though there has been an increase in the supply of fruits and vegetables, it is still below recommended values. In terms of utility, there are serious problems due to the estrangement of Caribbean populations from local diets and a shift to North American lifestyles with the consumption of processed foods, excess fats, and sugar. This has led to serious increases in diet-related diseases in the region. At the same time, while the nutritional profiles of most countries in the region is fair, there is evidence that pockets of malnutrition and micronutrient deficiencies is present in several countries (CFNI/FAO, 2007, p. 12). Rising cost of living; high rates of unemployment, poverty, and unequal income distribution; and inadequate social safety net programs in some countries impact entitlements (CFNI/FAO, 2007). In terms of the access dimension, there is, therefore, work to be done.

To address the food security threat in the Caribbean, the Caribbean Food and Nutrition Institute has called for multisectoral interventions aimed at reduction in fat and sugar intake, increase in intake of fruits and vegetables, reduction in consumption of animal products, and rationalization of food imports and policy-making, which consider all components of food security (Brocage, 2002). Food security policy should be tied to health and nutrition with attention to both over-nutrition and under-nutrition. This requires a philosophical shift from a conceptualization of food security as equivalent to food sufficiency (Brocage, 2002).

The past decade has seen a number of events and initiatives, which clearly indicate a certain level of urgency no doubt spurred by the recognition that food security is now a regional priority that can no longer be ignored. The World Food Summit set a goal of reducing global hunger by 50 percent by 2015. To this end the FAO established a Trust Fund for Food Security and Food Safety to be used to strengthen and sustain projects within the FAO Special Program for Food Security (SPFS). The increasing concerns about food security in the Caribbean prompted the CARIFORUM to ask the FAO to prepare a CARIFORUM Regional Special Programme for Food Security (CRSPFS) under the SPFS. The original deadline of 2007 was later extended to 2010. In 2002 the FAO and the CARICOM Secretariat collaborated and launched the US$26 Million Food Security Project. In April 2003, a US$5 million joint food security project was launched under the aegis of the CARICOM, the CARIFORUM, the FAO, and the Government of Italy. Then in December 2003, the Caribbean Food Security, Health and

Rural Poverty Program was launched by the CARIFORUM and the FAO aimed at ensuring food security, reducing poverty, and improving nutrition and health in the region (Caribbean Food Emporium, 2003). In July 2009, the University of the West Indies, Mona, hosted academics, agriculturalists, and leaders in a conference on Food Security and Agricultural Development in the Americas.

In addition, there have been a number of initiatives to build capacity of smallholder farmers throughout the region. For example, Farmer Field School Initiatives have been established in some countries. In Trinidad and Tobago, for instance, five farmer field schools were established, which provided technical advice and training for over a hundred farmers in areas such as irrigation, organic farming, and pest management In Dominica, a farmer field school provided training in organic vegetable cultivation, record keeping, and pest management (FAO, n.d., available at http://www.rlc.fao.org/progesp/pesa/caricom/images/Poster%209_Community%20food%2...). Farmers in Antigua have received training in vegetable cultivation and agrochemicals; farmers in Barbados have done soil moisture training; in Grenada trainings in soil and water testing have been conducted; packaging, mechanization, computer use, and other modernization techniques have been introduced to farmers in Jamaica, Belize, St. Lucia, and St. Vincent (FAO, n.d. available at http://www.rlc.fao.progesp/pesa/caricom/images/Poster%206_Country%20Specific%20...). Throughout the region, farmer and extension training initiatives are being implemented in critical areas including irrigation and water management, pest and disease management, record keeping, soil management, post-harvest losses, agrobusiness, organic farming, and marketing and distribution.

These are necessary activities to build production capacity among small-scale farmers, stimulate domestic production, and reduce the dependence on imported food through enhanced self-sufficiency. But we would argue that although they are necessary, they are not sufficient. Other important areas still not adequately addressed include the role of women, urban agriculture, and the role of local or traditional knowledge. A joint program by CFNI, CARICOM, and USDA to provide small agricultural grants has identified urban food production and community gardens for women as priority areas for funding (CFNI, USDA, CARICOM, 2003). It should be noted, however, that the Poverty Alleviation and Food Security Strategy of the Caribbean includes a Small Grants Program designed to promote innovative small agri-businesses was developed (CFNI, USDA and CARICOM, 2003). As we discuss later in this book, these are areas that can be fundamental to food security in the region based on research-based evidence from other parts of the world. It could be argued that there is incomplete and perhaps distorted knowledge about the extent of hunger in Caribbean populations. A fundamental problem is how food security is perceived. Typically there is an over-emphasis on food availability with not enough consideration of the other dimensions. Deep Ford (2003) argued that food policy should incorporate issues of nutrition, food security, food sovereignty, and the right to food and the prevention of chronic disease. But the lack of reliable information also makes accurate estimation difficult. This is compounded by the structure of Caribbean economies and the largely unaccounted for informal sector components.

PART II

Salient Issues in Food Production and Food Security in the Caribbean

CHAPTER 4

Decision-Making among Small-Scale Food Farmers in the Caribbean

4.1 Overview

We argue that in order to achieve greater food security and food self-sufficiency in the Caribbean, the capacity of the small-scale farming sector to produce more must be enhanced and that governments of the region must play a more central role in this regard. In order to be effective, governments and policy-makers must have a better sense of the environment in which small-scale farmers operate. This includes the decision environment in which decisions about what to plant, where, when, how, and how much are made. Decisions about marketing and distribution are also being made by farmers with implications for their livelihood and food security. Studying the decision-making framework of small-scale food farmers in the Caribbean—indeed in the tropics—is important as 90 percent of food in rural areas of the developing world is produced by such farmers (Josette Sheeran—World Food Program, 2009). An understanding of small-scale farmers' decision-making process can shed light on their activities and inform policy-making.

There are many factors that influence and inform the decisions farmers have to make on a daily basis about land use and other issues. The practices and decisions of small-scale food farmers in the tropics, including the Caribbean, have often come under scrutiny because there is still a general lack of understanding of the complex nature of decision-making among resource-poor farmers who invariably operate in challenging environments characterized by risk and uncertainty. With their low resource base, lack of institutional support, and production goals that are typically more nuanced than mere profit maximization, small-scale farmers' decisions can seem conservative, even irrational. This chapter looks at the decision-making process of small-scale food farmers in the Caribbean. We analyze the factors that influence decisions and articulate a conceptual model that summarizes the decision-making framework of Caribbean small-scale farmers.

4.2 Introduction

In Chapter 3 we provided an overview of the state of food security in the Caribbean. We suggested that the problem might have been underestimated until relatively recent times largely because the sensationalized incidents of famine and hunger, which have affected some other parts of the tropical developing world, are unknown in the region, with perhaps the exception of Haiti. We argued, however, that in light of certain factors, food security in the Caribbean should be considered precarious at best. Factors such as impact of climate change and the ubiquitous natural hazards, the decline in local food production in the region, the growing dependence on imported food, the estrangement of Caribbean populations from their traditional foods and diets and a movement toward North American style–fast food diets, and disruptions in the economic base and livelihoods in Caribbean rural communities constitute serious risks to regional food security. Our essential premise in this book is that improving food self-sufficiency and reducing dependence on imported food are big priorities for local agriculture and food security in the region.

Conventional wisdom places a great deal of emphasis on certain macro-factors as key elements that constrain food production in the Caribbean. These macro-factors include small farm size, under-utilization of land, farm fragmentation, insecure tenure, lack of marketing infrastructure, small resource base, and agricultural dualism marked by a small-scale food-producing sector to serve the domestic market operating alongside an export-oriented sector with which it competes—unsuccessfully—for scarce resources. We agree with Spence (1996) that analyses of these macro-factors are critical to enhancing our understanding of domestic food supply and hence food security, but that the decisions that farmers make are just as important. We would argue, however, that the macro-factors and the decisions that farmers make are not mutually exclusive. Farmers' decisions are, in fact, largely determined by these macro-factors or more precisely by farmers' perceptions and interpretations of how these factors operate and affect them. We would also argue that the impact of farmers' decisions on food production and food security in the Caribbean has not gotten much attention in the literature. There has been some work on the decision-making process of small-scale farmers (Beckford, 2002; Ryder, 1993) but little on the implication for food security (see Spence, 1996).

4.3 Agricultural Decision-Making among Small-Scale Domestic Food Producers in the Caribbean

On a daily basis, farmers are faced with numerous decisions about their cropping systems. Many of these are land-use decisions about what to plant, where, and how much, and whether it would be domestic or export production. In research from the Rio Grande Valley in eastern Jamaica, Davis-Morrison and Barker (1997) point to the flexible and dynamic nature of small-scale farmers' decision-making. They characterize this flexibility and adaptive orientation as a critical component of coping to the changing economic and environmental conditions related to farming in the area. These factors exert significant impact on the lives and livelihoods of farm families and so influence the choices farmers make in relation to cropping systems and land-use patterns over time (Davis-Morrison and Barker, 1997). The authors found a high degree of responsiveness of small-scale farmers to changes in the agro-ecology and argued that "contrary to conventional stereotypes many traditional small-scale farming systems in the Caribbean are dynamic and responsive to market opportunities" (p. 98). This perspective is important given the fact that the ability of small-scale farmers in the tropics to respond to changing conditions and adverse factors is not always recognized by policy-makers (Davis-Morrison and Barker, 1997; Hills, 1988; Chambers, 1983; Richards, 1985; Chambers et al., 1989).

Many decisions about land use are influenced by economic conditions. For example, in the upper Rio Grande Valley of Jamaica, the farming community responded to high international price for bananas under the preferential LOME trade agreement enjoyed by the Caribbean by investing heavily in the crop and using their best land to cultivate it. In the 1970s the bottom essentially fell out of the market as these farmers were unable to compete effectively with Windward Island and Central American bananas. In the early 1980s the government of Jamaica made production of nontraditional crops for export a priority. Most farmers in the area shifted land use to yam production, which became a lucrative cash crop. The destruction of the yam industry by Hurricane Gilbert in 1988 led to another shift in general land-use patterns in the area as dasheen, a lower maintenance crop that was less susceptible to hurricane winds and had export potential, became the main crop (Davis-Morrison and Barker, 1997). Evidence from St. Lucia and St. Vincent & The Grenadines in the Lesser Antilles reflect similar patterns. In St. Vincent's Mesopotamia Valley, some farmers responded to unfavorable market conditions, especially the removal of preferential trade arrangements in Europe and competition from Central American "dollar" bananas by turning banana farms into fruit orchards and small-scale plots for the cultivation of ground provisions and salad vegetables for household use and sale in local produce markets. In St. Lucia some farmers have responded to the challenges facing the banana industry by shifting land use to the cultivation of fruits and vegetables for the tourist market.

There is a tendency to favor export-oriented production to domestic-oriented production among small-scale Caribbean farmers (Spence, 1996). This export orientation bias supposedly has implications for food availability and consequently food security. It is true that small-scale farmers have always been involved in planting export crops. Even though export crops tend to be plantation crops, farms less than ten acres, which is the official definition of small farms, have always contributed significantly to the export of traditional export crops like sugar, coffee, cocoa, bananas, citrus, pimento, nutmeg, and tobacco. However, it should be remembered that small-scale farming in the Caribbean reflects an interesting hybridization of commercial and subsistence functions. This is especially true for small-scale food farmers. For commercially oriented farmers, crops with export potential do have more prestige because of their income-earning potential. It is unlikely, however, that this export market preference has any significant impact on food security. The main foods that now have export potential were initially grown for subsistence with surpluses being sold locally. This started during slavery on the provision grounds of slaves. In the 1960s some export of food crops started opening up under a new export trade category called nontraditional exports. Exports of these crops, however, are but a small proportion of the total production. For example, in Jamaica the most significant food crop exported is yams. The most popular subspecies of yam exported is yellow yam with Jamaica owning a monopoly of the world yellow yam market. Despite this the country exports just 5–6 percent of its total yellow yam output. The export of yellow yams has no impact on yellow yam availability in the domestic market and consequently does not adversely affect food security. Furthermore, the traditional food crops that are exported today are also staples in local diet and local cuisine and are almost always found in abundance in local produce markets except in extenuating circumstances such as after natural disasters. The bottom line is that small-scale food farmers do not participate in the export market to the extent that it takes away resources or attention from producing for the local market. In fact, by exporting some of their produce, they reduce the quantity available for the domestic market, which often experiences over-supply of local foods with adverse implications in terms of prices.

A major land-use factor with serious implications for food production and local food security is the under-utilization of land on small-scale domestic food farms. One of the main characteristics of food farming in the Caribbean is the very high incidence of farm land that is not in production at any given time. In a study in the Rio Grande River Valley of Jamaica, it was found that of a total of 352.4 acres controlled by the study sample, 62 acres were under cultivation and another 82 acres were used for pasture, which means that over 50 percent of the available arable land was not being used or was being under-used (Meikle, 1998). In studies in Cascade, Hanover, and Sawyers, Trelawny, both in Jamaica, it was found that 52 and 44 percent of farm land, respectively, was unused (Meikle, 1992). Small-scale domestic food producers operate farms between 0.5 and 4.0 acres in size and so unused land takes on even greater importance.

There are several factors that could explain this pattern of high unused land on small farms. One explanation is that typically some of the land is classified as being rested, that is laying in fallow. The most common explanation farmers give for unused land, however, is that they do not have the capital resources to develop more of the land at their disposal. While some farmers rest land as a deliberate agronomic technique, the vast majority of unused land is actually idle. Some fallow occurs by default rather than as an agricultural practice, and this has to be understood when examining this issue. Typically farmers' decision to rest land or leave it idle is determined by a lack of the resources to clear more land, hire labor, and afford farm inputs. Marketing considerations also play a role and farmers often respond to adverse market conditions by taking land out of production and scaling back cultivation. It is important to point out that small-scale food farmers have adapted to small farm size with strategies such as intensification of farming and mixed cropping.

4.4 A Model of the Decision-Making Framework of Small-Scale Caribbean Farmers

Figure 4.1 summarizes the complex set of interrelated factors that inform the decision-making process of small-scale farmers in the Caribbean. The model is derived from the synthesis of research on small-scale farming conducted over the past 30 years. The first point to note is that human decision-making is an enormously complex process and the results of research in this area must be treated with great care when causal relationships are being established and conclusions drawn. This is because objective observation of human behavior is imprecise. Studies may reveal valuable insights into the decision-making process, but results rarely yield universal applicability. Like the model presented here, they should not be interpreted as foolproof models of reality but as general frameworks for understanding and analysis,

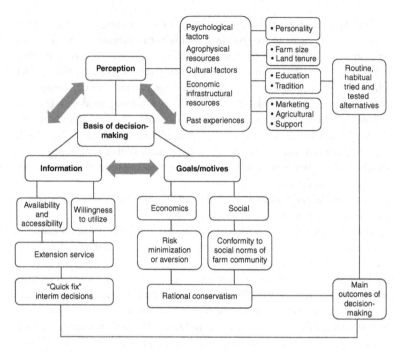

Figure 4.1 Small-scale famers' decision-making in the Caribbean

which represent works in progress rather than static explanations. Furthermore, they should be interpreted as situational and contextual, spatially and otherwise.

Found (1971) emphasized the difficulty of observing human behavior objectively. Quite often, farmers are unclear of the basis of a decision and may be unconsciously or subconsciously affected by certain factors. It is thus possible that under seemingly identical circumstances, decisions vary widely among and between farmers. Ilbery (1985) suggests that a degree of randomness is a common characteristic of farmers' decision-making.

4.4.1 Factors That Influence Small-Scale Farmers' Decision-Making

The decision-making behavior of small-scale food farmers in the Caribbean reflects a number of goals, with economic security being primary. A critical component of the decision-making process is the extent of knowledge and information available to farmers. This is a major component of most behavior theories and decision-making models. According to Found (1971), when forming expectations, farmers proceed by a process of elimination.

It is fundamental to distinguish between two types of environment defined by the amount of information available and the impact on farmers' behavior (Found, 1971; Ilbery, 1985). The information that is available to farmers is called the *decision-making environment,* while the *extended* or *real* environment encompasses the complete set of information available (Found, 1971; Ilbery, 1985). The two are quite different as most farmers do not consider it necessary to learn everything about the extended environment. Therefore, the level of conscious decision-making in most cases depends on the decision-making environment. Supposedly, more educated farmers will acquire more information and will have a better impression of the extended environment, and their behavior may reflect this (Found, 1971). In the Caribbean context the difference between the real or extended environment and the decision environment is significant because small-scale food farming attracts mainly people with low levels of educational attainment.

The decision-making environment is also profoundly affected by farmers' perceptions and assessment of the benefits and costs of a particular course of action. This is fundamental because, as Ilbery points out, "it is the way in which a farmer perceives the world which is important, rather than what is actually there" (1985, p. 31). The importance of perception is based on its influence on the learning process through which images of the decision environment are formulated (Soarinen, 1969). In studying decision-making, therefore, due consideration should be given to farmers' own images of their environment. In the Caribbean, perception is a major influence on the decisions of small-scale farmers. Perception is shaped by a range of factors that are captured in figure 4.1, including economic, cultural, social and psychological, physical, and family needs and, very importantly, the individual and collective experiences of farmers.

Small-scale food farmers' decision-making also seem to have an element that we describe here as a *community effect.* It is generally felt that the decision environment is strongly influenced by the nature of the decision-maker, and given that human beings are basically unique and distinct, it would be expected that this would be reflected in the decision-making of farmers. While differences are manifested among farmers, it has also been observed that there is a high degree of convergence in some farming communities in relation to crucial issues like land use, resource management, marketing, and attitude toward agricultural innovations (Beckford, 2002).

In analyzing the decision-making process of small-scale farmers, we addressed the issue of the goals of farmers. Specific consideration was given to the question of whether or not farmers' goal was to optimize decisions and maximize profits. The principle of optimality is founded in a classical economic assumption that "economic man is rational"; hence when faced with certain situations, the "economic man" will be motivated to maximize benefits (e.g., profits). The concept of economic man implies that decision-makers have perfect knowledge and complete information upon which to base decisions and make choices. Not surprisingly, many writers have argued that optimality is unattainable in the real world (see Found, 1971; Ilbery, 1985; Simon, 1957; Blunden, 1977). Ilbery (1985) suggests several points about decision-making, which he argues lend more weight to the argument that farmers cannot make optimal decisions. One is that the extent of the differences between the decision and extended environments depends on the ability and motivation of farmers to search for information (both of which are usually low). Second, the decision environment varies spatially, and third, social and psychological factors are important in the decision-making process. It is in this context that *satisfaction* is being offered as an alternative motive to optimization. The *satisficer* concept (Simon, 1957) is attractive for obvious reasons. Not only is it logical, but it is safe in that it lacks the extremities of the optimality premise (Van Den Ban and Hawkins, 1988). In the context of the Caribbean, the satisficer concept is a more accurate explanation of the motives of small-scale farmers as there are a plethora of limitations and constraints that hinder optimum outcomes.

Herbert Simon (1957) conceptualized the principle of decision-makers being satisficers and introduced the principle of *bounded* or subjective rationality as opposed to "omniscient" rationality (Ilbery, 1985) or "traditional or objective rationality" (Found, 1971). Objective rationality, he argued, was beyond the scope or capabilities of real people for two reasons. First, information is not a free good and must be searched for within the constraints of time, finances, and the like, and second, decision-makers have only limited capabilities to process the information that has been acquired.

Simon's arguments became the basis of later decision-making models and are relevant in the Caribbean contexts. In making decisions, small-scale farmers in the Caribbean prefer to fit into an already established pattern. Making substantial changes exposes them to the possibility of making poor decisions—a situation they like to avoid. They therefore prefer to adopt systems and practices that have been tried and proven by themselves, their neighbors and their elders, rather than experiment with new practices that might adversely impact the well-being of their families. But while the satisficer concept more aptly describes farmers' motivations than the optimization concept, it is also clear that avoidance of risky choices is a fundamental element of farmers' decision-making. Small-scale food farmers in the Caribbean are clearly more concerned with reducing losses and minimizing risks than maximizing profits. It should be noted that there is a relationship between profit maximization and risks. It could be argued that options that potentially maximize profits will also require more risk taking, other things being equal.

It might also be the case that focusing on risk minimization more often than not produces a satisfactory outcome. What is unquestionable, however, is that Caribbean small-scale farmers' decision-making is fundamentally motivated by a desire to minimize risk (Beckford, 2002; Spence, 1996).

It should also be said that the decision environment is very closely related to culture and practice. Systems and strategies that have stood the test of time, expected returns on investments, values and beliefs, and basic motivations are strongly influenced by social orientations of the farming community. This perhaps explains the observed community effect or group behavior that was discussed earlier.

On the evidence of the discussion up to this point, it can be concluded that generally speaking, the explanation of the decision-making process of small-scale farmers in the Caribbean contains a large element of convergence with the existing literature on the subject. There are, however, several critical characteristics that either are not considered in the general literature on the subject or represent divergence from general theories.

The first of these is referred to here as the *quick fix syndrome* (Beckford, 2002) and describes the observed tendency of farmers to make decisions that offer quick and, in many cases, temporary solutions to problems being experienced by individual farmers and the farm community. The existing decision-making literature tends to focus on the question of how realistic it is to assume that decision-makers are perfectly rational and seek to optimize their decisions. Many writers have concluded that perfect rationality is impossible and offer the more realistic view of the satisficer motive discussed earlier (see Soarinen, 1969; Simon, 1957; Found, 1971; Blunden, 1977; Ilbery, 1985; Van Den Ban and Hawkins, 1988).

It has been argued that in the search for a solution to a problem, the first satisfactory solution is often chosen (Ilbery, 1985). We do not contradict this opinion, but would argue that among Caribbean small-scale farmers, this first satisfactory alternative is generally accepted initially on a temporary basis as a "quick fix" or interim measure. In the literature, the permanency of the first satisfactory solution is implied, but in the Caribbean, many decisions made by small-scale farmers represent "stop gap" measures, while the search continues for a solution that will yield a greater level of satisfaction. If the quick-fix measure works fairly well (in the context of farmers' goals) and yields a high degree of satisfaction (which might still be short of the desired satisfaction), the search will not be intense. If, on the other hand, the quick-fix measure yields little satisfaction for the farmer, the search process will proceed with more urgency.

An example can be used to illustrate this point. Yam cultivation in the prolific producing areas in central Jamaica is dependent upon the use of sticks for yam vines to climb. Farmers have responded to severe challenges in obtaining this critical input by turning increasingly to the use of two species of plants—bamboo and Brazil Maca, which are not traditionally preferred stakes—and recycling yam sticks. Recycling yam sticks refers to the practice of combining old sticks and reusing them the next cropping cycle or season. These two responses started out as short-term remedies but have become features of the cultural ecology of yam cultivation as the yam stick problems persist and deepen and the stopgap measures work satisfactorily. While they practice these responses, however, they continue to hope for the ideal solution or a more satisfactory solution. The word *hope* is used instead of *search* because there did not seem to be an active search in progress for a more permanent or satisfactory solution. From farmers' description of the two responses, however, it can be deduced that some more satisfactory option would be desirable. Furthermore, a "problem" with an existing methodology often surfaces only when new ideas or techniques that represent points for comparison emerge. In other words, a satisfactory way of doing things can suddenly become unsatisfactory when we are introduced to other seemingly superior ways of doing the same thing.

The ideal and sustainable solution to the yam stick problem from the perspective of the farmers rests in finding an affordable and abundant source of good-quality yam sticks with a life span of 3–5 years or more. The response discussed here happens to offer a fairly high degree of satisfaction, but the yam stick problem was identified by farmers as their biggest challenge and so a more sustainable option would be attractive to farmers. The point we are making is that quick-fix measures can continue for long term and take on a look of permanence as long as they make economic sense and so long as better solutions remain elusive. For example, the recycling of yam stick, which was a quick fix to shortage of good-quality yam sticks, has become a feature of commercial yam cultivation in Jamaica since the late 1980s or the early 1990s.

A second point of divergence relates to the question of initiative on the part of the farmers. The responses to the yam stick problem by farmers, especially the recycling of sticks, speaks to their ability to be adaptive, creative, and resolute. It would seem, though, that when it comes to making certain important decisions, small-scale farmers need assistance and assurance from "experts." This is so even where they acknowledge that a particular course of action or land-use decision is rational and profitable. More specifically, they clearly operate more rationally in the decision environment when assistance and assurance are forthcoming from people they consider to be experts, such as extension officers. The more complex and technical the issue is, the more input is needed from experts. The reliance on extension services in such a case is understandable given the modest educational level among the farmers in question. Extension officers can therefore play a big role in the decision-making of small-scale farmers by helping them to diagnose problems, synthesize information, and appraise possible solutions.

A third point relates to the divergence between the farmers and scientists about the value of a course of action. This point is not given a lot of prominence in the literature, but studies in the Caribbean have pointed to differences in how a new idea or system is perceived by farmers compared to policy-makers and scientists who create them (Collymore, 1985, 1986; Spence, 1996; Beckford, 2002). This is especially so in relation to decisions about adoption or non-adoption of agricultural innovations that we discuss in Chapter 5. One reason for this divergence of opinions is that farmers and scientists view the same problem and its solutions through different mental windows with the entire goals of the farmer unknown to or misunderstood by the scientists. Writing about people's reactions to environmental hazards, Burton et al. (1978) suggest that people living in hazard-prone areas have different views of the hazard than an expert studying the phenomenon. They are not necessarily wrong in their appraisals; they focus on different characteristics and given their particular needs may actually arrive at more accurate or useful appraisals than the scientists or "experts."

A final point of interest relates to the inherently conservative nature of farmers, which is often assumed. The tendency of small-scale farmers to be cautious is sometimes seen as a major obstacle to the growth of agricultural production in the Caribbean. Small-scale farmers operate in an environment that is rife with risks and uncertainties. Natural phenomena such as climatic hazards, volcanoes, diseases, and pests combine with other factors such as unfavorable market conditions, lack of access to resources, and insecurity of land tenure, among others, to create a challenging agro-ecological environment. Small farmers generally have little economic security in terms of savings, no crop insurance, and a very small resource base, and with the welfare of their households at stake, their management strategies are more often than not geared toward minimizing risks. This might indeed be justifiably described as conservative, but is not only understandable but also rational (Beckford, 2002). Evidence from research among Caribbean small-scale farming communities in the Caribbean indicates that farmers' cautious safety first approach is based on assessments of the net worth of a particular course of action in the context of their special circumstances. Beckford (2002) has interpreted this approach as "rational conservatism" motivated by the goals of risk minimization or aversion; in an atmosphere of uncertainty and in the context of economic insecurity and a small resource base, the consequences of failure are great (Wigley, 1988). The desire to minimize risk makes good sense in this context. Hudson (1981) suggests that the circumstances under which small-scale farmers operate require them to adopt a strategy geared toward safety. The essence of the small farmers' problems is reflected in the words of Benito (1976, p. 8), who describes small-farmers' behavior as "survival type of algorithm . . . as safety first rules."

4.5 Summary

Contrary to the common perception, decision-making among small-scale food farmers in the Caribbean, as is the case elsewhere in the tropics, is quite dynamic and adaptive with decisions constantly being re-assessed and re-appraised. Adjustments are common in response to these re-assessments. Small-scale farmers' safety-first mantra is the product of their lived experiences and the lack of institutional support and relief. We have seen elsewhere in this book how the dualistic structure of agriculture in the region and certain structural biases against the small-scale farming sector act as serious constraints and disincentives to increased productivity. Governments should recognize that they can influence farm-level agricultural decisions. Decisions that are progressive and will lead to modernization, increased efficiency, greater productivity, improved food self-sufficiency, and more environmentally sustainable agricultural production will require the dismantling of the structural biases against the small-farming sector, provision of a level playing field to compete with food imports, adequate support and relief in the face of natural hazards, and the removal of the many constraints to production.

A critical aspect of decision-making with serious implications for local food production is related to agricultural modernization. The next chapter analyzes critical issues around farmers' decisions specifically in the context of agricultural innovations.

CHAPTER 5

Factors Influencing Innovation Adoption among Small-Scale Farmers in the Caribbean

5.1 Overview

There is a tendency to stereotype Caribbean small-scale farmers and tropical small-scale farmers, more generally, as being resistant to change and innovations. This negative stereotype assumes that the decision of farmers to not adopt an innovation has no rational basis and is due mainly to apathy and a general resistance to change. But the situation is far more complex and small-scale farmers can be innovative. It might even be argued that, generally speaking, the decisions of farmers not to adopt an innovation, to postpone adopting an innovation, or to discontinue the use of a new practice is based on rational considerations in the circumstances of the farmers (Beckford, 2000, 2002; Collymore, 1984, 1985; Spence, 1989, 1996; Davis-Morrison and Barker, 1997). Furthermore, there are ample examples that small-scale farmers are engaged in farm-level experimentation and innovation on a daily basis and that this is critical to their survival.

Research on the subject indicates that the way farmers make decisions about agricultural innovations is a complicated and misunderstood process and observers often mistakenly attribute non-adoption to an inherent unwillingness to accept change by farmers (Beckford, 2002; Ryder, 1993; Spence, 1989). We argue that small-scale farmers can be and often are innovative, and are adaptive, flexible, and dynamic, constantly reappraising and reassessing their decisions. Farmers are not diametrically opposed to innovations. They assess them within a particular context, and if they find the innovations to be useful, and can afford them, they will adopt them.

But farmers must be convinced of the value of an innovation. Also the level of risk involved in adopting any new idea must be tolerable to the farmer. Tolerance is subjective and related to the circumstances farmers find themselves in at any given time. Generally, this threshold of risk is considered to be low with Caribbean small-scale farmers, who, as we discussed in the preceding chapter, are known to be risk minimizers. Given their low resource base and lack of institutional support, this is understandable. The choices they make are influenced by many factors, but particularly those of an economic and physical nature, which are usually exogenous. A critical component of Caribbean small-scale farmers' response to modernization packages is their perception of an innovation or a choice. It should be noted, too, that farmers' views about an innovation often differs from those of scientists or the innovators themselves and agricultural authorities. Farmers and scientists also bring different frames of reference for analyzing situations.

In this chapter we analyze the behavior of small-scale farmers in the Caribbean in relation to modernization efforts and agricultural innovation adoption.

5.2 Introduction

The global food crisis of the 1970s and 1980s shone the spotlight on the capacity of small-scale food production systems in the tropics and subtropics to produce food and increase food self-sufficiency and food security. Following the relative successes of the Green Revolution technology in the 1960s, it was thought that modernization of traditional food systems in the developing areas of Africa, Asia, Latin America, and the Caribbean would provide an impetus for increased domestic food production. In Jamaica, food shortages in the 1980s and tensions between the island government and the government of the United States underscored the risks of depending on imported food and prompted a food strategy built around increasing food self-sufficiency and eating locally grown foods. But the goal of modernization of domestic agriculture had always been tenuous—not just in the Caribbean but throughout the developing world. The literature is replete with examples of agricultural modernization interventions that had failed miserably or achieved insignificant success. The literature on small-scale farming in the Caribbean is also characterized by a concern over the high degree of failure in agricultural intervention strategies aimed at stimulating development and improving small farming performance (Demas, 1965; Nurse, 1970; Barker et al., 1983; Collymore, 1984, 1986; Spence, 1989).

Collymore (1985) attributes this lack of success to inadequate consideration of the environment in which farming decisions are made. According to

Spence (1989), the unidirectional flow of communication from agricultural innovators and planners is a reason for failure of modernization in agriculture as this failed to consider the complexities of the environment in which small-scale farmers typically operate. Spence argued that "[t]he attitude of traditional farmers to agricultural modernization is largely determined by the way they interpret and analyze factors of change" (p. 220). In his view, this underscored the need for agricultural planners and innovators to seriously consider the perceptions and attitudes of the target population when modernization plans were being developed.

Collymore (1986) also observed the general failure of agricultural modernization efforts in the small-scale food production sector. He attributed this partly to the inadequate attention to the contexts in which small-scale farming decisions are typically made. He argued that agricultural planners ignoring the nuances of small-scale farmers' resource management decisions applied a "cold rationalism" to their approach. This is in contrast to the "subjective rationality" of small-scale farmers (1986). Convincing small-scale farmers to accept modernization initiatives will continue to fail unless those who present modernization and innovation packages modify their cold rationality. This can be done only by understanding their decision-making processes (Collymore, 1986, 1985).

Collymore's work in St. Vincent and Spence's work in Jamaica are substantiated by the work of Ryder in the Dominican Republic (Ryder, 1993). Like Spence and Collymore, Ryder points to the flexible nature of decision-making and the importance of communication between agricultural innovators and the end user small-scale farming population. Wigley (1988) noted the uniqueness of tropical small-scale farming systems, emphasizing an important distinction between such systems and commercial farmers in developing countries. The former are influenced by factors that are more important than market forces, and hence seemingly backward decisions often have a rational basis in the decision-making schema of small-scale farmers (Wigley, 1988).

5.3 Factors Influencing Farmers' Adoption Decisions

Reporting on research in the yam-growing region of Jamaica, Beckford (2002) used the point score analysis to identify five categories of factors that influenced the attitudes and decisions of small-scale yam farmers in relation to the adoption of two specific innovations (see Boxes 5.1 and 5.2). These were as follows:

A. Personal factors
B. Sociocultural factors

C. Economic factors
D. Institutional factors
E. Environmental factors

Box 5.1 Minisett technology

The minisett system of yam cultivation is a technological package comprised of several components that distinguishes it from the traditional yam-growing system. The main ones include smaller planting setts, continuous instead of individual mounds, use of plastic mulch, use of chemicals for treating setts, the non-use of yam sticks, and production of smaller, uniformly shaped tubers. Minisett is based on the rationale that all parts of the yam tuber (that is, the head, middle, and tail) can sprout and produce yams. This contrasts with the traditional system where only head setts are used, yams are planted in individual mounds, and a 12-feet pole called a yam stick provides support for the climbing yam vines. The system was developed in Nigeria, and a minisett is defined as a sett that is less than a quarter of the normal planting piece (Okoli et al., 1992). It involves the cutting of mother tubers into smaller panting setts. The setts are treated with chemicals to prevent rotting before being put in a nursery from which they are transplanted in outdoor fields.

Minisett was introduced in Jamaica in 1985 as part of a strategy to improve the husbandry practices of yam cultivation, thereby enhancing the competitiveness of Jamaican yams (Campbell-Chin Sue et al., 1995). Widespread adoption would mitigate some environmental problems with the traditional yam stick method such as accelerated erosion of soil on steep slopes and the removal of trees from hillsides to be used as yam sticks. The system is touted as having several advantages over the traditional cultivation method. These include increased yields, better-quality planting stock, more marketable produce, reduced labor requirements, more environment friendly, and facilitation of mechanization of key activities.

Despite these stated advantages, the minisett system has had very little success among small-scale farmers in the main yam growing region of the country. Farmers have remained resolute in their assessment that the minisett system cannot improve their fortunes. More than two and a half decades after its introduction, research still indicates a very high level of disaffection and apathy toward the minisett system, which they derisively call "plastic yam." Farmers contend that the tubers produced are too small and hence not marketable, the system is not suited for the steep slopes that they cultivate, they do not understand the system, and due to slope orientation, the system would not work on some of their lands. That said, it should be noted that a recent study showed a shift in attitude with more farmers reporting that they would be prepared to consider the minisett system (Beckford et al., 2011).

Box 5.2 Agroforestry for yam stick production

In 1989, the Food and Agricultural Organization (FAO) conducted a national forestry review of Jamaica, under an FAO-coordinated Tropical Forestry Action Plan. The result was the preparation of the National Forestry Action Plan (NFAP) for Jamaica a year later (Evans, 1994). The primary objective was to generate increased benefits and better livelihoods for rural communities. The strategy was to introduce a wide range of agroforestry activities in farming communities in rural Jamaica. The yam-growing area of central Jamaica was identified by the FAO as an area deserving of urgent initial attention. It was estimated that around 8,000 farmers were involved in commercial yam cultivation in the region and that they required in excess of 6,000,000 yam sticks annually. The FAO concluded that years of yam stick harvesting from the wild was causing forest degradation (Evans, 1994). To address the high demand for yam sticks, the FAO developed a project called "Agroforestry Activities in Yam Growing Areas." The project was concentrated in the main yam-growing region of the country and included prolific yam-growing communities in south Trelawny, north Manchester, and north Clarendon. The principal demand for wood in this region centers around the demand for yam sticks, which was traditionally met by harvesting sticks from local forests and woodlands. Decades of exploitation led to depletion of local sources and precipitated the growth of an informal commercial trade in yam sticks. The decreasing supply of good-quality sticks and the increasing cost of the input constitute serious threats to the long-term viability of yam cultivation in central Jamaica.

Three main agroforestry initiatives were implemented to address the situation:

1. Live yam stick system where yam sticks are managed to remain green and alive year after year without increasing in size and reducing the need for new sticks

2. Live tree support system where trees were established in yam fields to serve as living support for yam plants

3. High-density plantations for yam stick utilizing fast-growing tree species

There was broad-based support among farmers for initiatives 1 and 3. Yet adoption level among farmers in terms of the live stick system was very low. In terms of the first initiative, most farmers did not have the resources to establish trees on their lands to produce yam sticks or they did not have sufficient land to engage in this activity. In regard to the third initiative, more than 95 percent of farmers reported that that they would be willing to purchase sticks from yam stick plantations if they were good quality and affordable. Unfortunately no such plantations were actually established by the FAO project and there has been no follow-up on the part of the authorities.

The factor scores indicated that economic, sociocultural, and, to a lesser extent, institutional factors were very important in the attitude of farmers toward these two specific innovations. It was found that personal factors, which would include personal preference and resistance to change, were not major factors that influenced the decisions of farmers whether to adopt innovations or not. This is an interesting finding as there is a tendency to explain farmers' rejection of modernization schemes in terms of some innate resistance to change. Perhaps surprisingly, environmental factors were also less important. This finding actually reflects the resilience and adaptability of small-scale farmers. Research in Caribbean countries such as Guyana (see Semple, 1996), St. Vincent (see Collymore, 1984, 1985), Jamaica (Campbell, 2011; Campbell and Beckford, 2010; Beckford and Bailey, 2009; Beckford, Barker, and Bailey, 2007; Davis-Morrison, 1998; Meikle, 1998; Spence, 1989), Dominican Republic (Ryder, 1993), Grenada (Brierley, 1976, 1991), and Montserrat (Thomasson, 1994), among others, has consistently pointed to the ability of farmers to manipulate or strategically exploit their physical environs to eke out an existence. Reporting on his research in St. Vincent, Collymore (1986) observed that although farmers are heavily dependent on the physical environment, they were "not pawns in the physical environmental forces surrounding them" (p. 94). Farmers react to their environment based on their perceptions and analyses, which, because of their dynamic nature, are constantly evolving based on the observations and experimentations of individual farmers as well as the collective knowledge of the wider farming community.

5.4 Correlates of Adoption of Agricultural Innovations among Small-Scale Farmers in the Caribbean

Ilbery uses the term "correlates of adoption" to refer to various interrelated factors at work in the decision-making process of farmers about agricultural innovations and their ultimate utility to them. The various and interrelated factors include personal, sociological, psychological, situational, and infrastructural issues. The model illustrated in figure 5.1 synthesizes the correlates of innovation adoption among small-scale food farmers in the Caribbean. The model indicates that there is a complex web of processes at work in the decision-making environment. If a high level of innovativeness is to be achieved, many of the associated factors need to be interrelated (Ilbery, 1985).

There is a tendency to depict non-adoption as the result of lack of education, closed-mindedness, and a stubborn unwillingness to change even when confronted with clearly superior options. Research evidence from the Caribbean provide a different picture where decisions about innovation adoption can, in fact, be viewed as considered and educated (see Collymore, 1985, 1984; Wigley, 1988; Spence, 1996; Davis-Morrison, 1998; Beckford, 2002; Barker and Beckford, 2006).

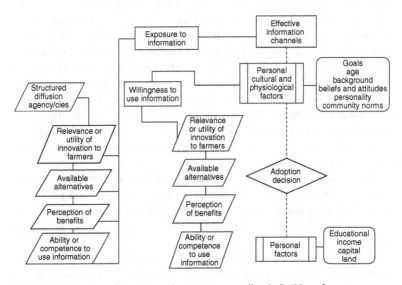

Figure 5.1 Correlates of innovation adoption among small-scale Caribbean farmers

According to Hagerstrand (1967), the diffusion of any innovation is essentially a spatial process. He conceptualized the adoption of innovation as the result of a learning process. This implies that factors related to the *effective flow of information* are critical. It implies too that the identification of the spatial characteristics of the target area and resistances/barriers to adoption is a fundamental step in examining the process of innovation diffusion. The role of effective information flow as a correlate of innovation adoption may be analyzed with reference to the diffusion of minisett technology for yam cultivation (Box 5.1) and agroforestry for yam stick production (Box 5.2). In both cases the lack of effective information to farmers seems to have played a major part in the low rates of adoption. In the case of minisett, in particular, a major problem was the absence of an effective strategy for educating the farming public about the benefits of the system. The amount, type, and quality of information available to farmers are critical to the successful diffusion of an innovation as these determine farmers' perceptions of and consequently their attitude toward the innovation.

Perception is a fundamental correlate of innovation adoption in the Caribbean, and farmers' decisions are significantly affected by their perceptions and assessments of the pros and cons of adopting an innovation. Ilbery (1985) posited that the way a farmer sees the world is more important than what is really there. Perception is important because it significantly affects the learning process through which images of the decision environment is formulated (Beckford, 2002). Decision-makers rarely have full information, which makes decision-making a less than fool-proof endeavor under the best of circumstances. It is no surprise, therefore, that among small-scale farmers, there is a close relationship between perceptions and information and communication. The trouble is that quite often the sources of information upon which farmers rely are unreliable. Their perspectives may be based upon hearsay and vicarious experience rather than on personal or first-hand knowledge. This was the case, for example, in the attitude of yam farmers about the merits of minisett, a new system for cultivation of yams (see Box 5.1). In this case, farmers had come to the conclusion that this new system did not represent improvement.

Despite official views to the contrary, farmers felt that this system produced yams that were less marketable than those produced from the traditional cultivation system. Part of the problem here was the lack of extension services and the lack of a coherent flow of information about the system to farmers.

Yet access to more information in and of itself might not have led to a different view of the system as there are typically many other factors at work in the decision environment. In addition to economic, sociocultural, and environmental factors, Jones (1963) posits that the characteristics of farmers may determine how they actually use information they acquire. There is still no overwhelming consensus about the extent to which adoption patterns are determined by patterns of information flow and the extent to which they are determined by the characteristics of adopters (Blaikie, 1978). However, the perceptions and evaluations of farmers are important as they are the persons who determine the utility of an innovation (Beckford, 2002).

This brings us to another significant factor that informs farmers' response to agricultural innovations—their assessment of its value or its utility to them. In the minisett example discussed earlier in text, a number of purported advantageous attributes of minisett were presented. However, there is divergence between the appraisals of farmers and agricultural planners. Farmers obviously did not buy into the espoused benefits of the system, hence their rejection. In the eyes of the experts and agricultural policy-makers, this rejection may reflect backwardness and an unwillingness to accept change. For the most part, however, we consider the reasons given for not adopting minisett to be logical and rational. For example, farmers cited issues such as terrain constraints in terms of its suitability for hillside agriculture, unsuitability for some slopes in terms of exposure to sunlight, its complex nature, and small tuber size and marketability as deterrents to them practicing minisett.

Not surprisingly, economic security of the farm family was the most critical consideration in farmers' decision-making about agricultural innovations. Small-scale farmers in the Caribbean will rarely engage in economic behavior, which in their estimation jeopardizes the security of their family. In this context farmers report concerns specific to their ability to put adequate food on the table at all times, security of the homestead, and their ability to fund the education of their children and provide other financial assistance to them. As we established in the chapter 4, farmers' decision-making occurs in agro-ecological contexts rife with risks and uncertainty. Given the low resource base that typically characterizes their operations, there is often very little margin for error. Small-scale tropical farmers are considered to have a high degree of aversion to risks in the business environment. As risk minimizers, farmers would thus be unwilling to go for options where they have a lot to lose.

Much is often made of the relationship between an individual's adoption index and his or her social characteristics (Jones, 1975; Van Den Ban and Hawkins, 1988). Age and reading ability are often assumed to correlate with the adoption of decision in that younger and more educated persons are more likely to be early adopters than older, less educated people who are thought to be more resistant to change. This is not always the case in the Caribbean (Beckford, 2002). This is supported by Van Den Ban and Hawkins (1988), who found that reading ability had very low correlation with the adoption index. In relation to age, they argue that, contrary to popular opinion, in half the studies conducted on the subject, no relationship was found between age and adoption of innovations. Furthermore, only a third of studies show that younger people were more inclined to be innovative than older people. This contradicts the view of Jones (1975), who postulated a positive relationship between education, information, and adoption. He also pointed out the tendency for older farmers to be less keen on innovation adoption.

In the Caribbean there is a correlation between the *characteristics of an innovation* and its rate of adoption:

(i)   The advantages of the innovation vis-a-vis the status quo
(ii) The extent to which the innovation is compatible with the sociocultural values, beliefs, and traditional ideas of the farming community and the needs of farmers in that community
(iii) The ease with which the innovation can be applied

Farmers' attitude to minisett indicates that the system is unattractive because they think that it cannot help them (i.e., relative advantage), it does not fit into the cultural orientation of producing large tubers (compatibility), and it is highly technical as it involves several steps (i.e., complexity).

The literature on innovation adoption places a lot of attention on the characteristics of innovation users as a correlate of innovation adoption or rejection (Rogers, 1995). According to O'Loughlin et al. (1998), the decision on innovation adoption is significantly affected by a combination of factors including local perceptions of the innovation as an alternative to the status quo and the respective influence of the advocates of the competing positions. This raises the question of the general approach to innovation diffusion, which more often than not is characterized by a top–down approach that presumes that new ideas are intrinsically valuable and superior to existing methodologies and also appropriate to the local context (Chambers, 1997, 1983). Furthermore, it is presumed that this value and superiority should be self-evident. As we have seen time and again in the Caribbean and other tropical small-scale farming systems, however, this does not always obtain.

The role of *opinion leaders* in the farm community is an important adoption correlate to be considered. Opinion leaders are influential farmers in the community whose views and practices are respected and emulated. They are usually successful farmers, relatively progressive, and are considered to be "big" farmers in terms of the size of their holdings or operations. They become opinion leaders because in addition to the foregoing attributes, they are willing to share ideas with other farmers and help them solve problems and make important decisions. They also tend to have more contact with extension officers and the seats of power and influence. In the yam stick studies in Jamaica, the views of opinion leaders were reflected by the general farming community, thus underscoring their influence.

The role of opinion leaders has been established in the general literature on the subject. Van Den Ban and Hawkins (1988) identify five central functions of opinion leaders. These include passing on outside information, interpreting information from their own perspective, setting an example for others, giving approval or disapproval of changes planned or implemented, and exerting influence in changing group norms. Lin Nan (1973) cautioned that not all opinion leaders do all of these things. Some provide information early in the adoption process, while others legitimize the decision to adopt or reject an innovation. Zawisza and Pilarska (2005) in research from Poland found that opinion leaders own farms that have more potential than other persons, were better educated than others, showed more interest in obtaining information, and felt more need to try new things on their farms.

One issue that is still to be resolved is what should be the focus of innovation diffusion analysis. Should it be skewed toward the demand side—that is, focus on receivers? Or should the focus be on the supply side—that is, attention on the diffusion agency? We would argue that only a strategy focusing on both aspects will be successful. Physical, economic, psychological, personal, and sociocultural factors of farmers play a key role in their response to innovations. But there is also evidence that a poorly designed diffusion strategy characterized by inadequate information flow is also detrimental to the diffusion process and adoption rates among farmers.

5.5 Summary

If the Caribbean is to achieve sustainable food security, the region must become more self-sufficient in food supply. This will require modernization of the small-scale farming sector, which remains the backbone of local agricultural production. A better understanding of the factors that influence farmers responses to agricultural innovations is central to modernizing the sector and increasing its capacity to produce more and better-quality food.

Small-scale farmers will adopt agricultural innovations if they are convinced of their utility and value, and if they can afford them. This can take an extended period of time as farmers' views about the value of an innovation tend to differ greatly from those of experts. Indeed, innovators and scientists are often more concerned about the perceived intrinsic value of their creations, while farmers naturally are concerned only about its usefulness and whether it represents an improvement over the status quo. A particular innovation may have potential value, but this potential may not be realized under normal farm conditions—a situation that renders it useless to farmers.

Despite rationality and innovativeness among small-scale farmers, traditional beliefs and practices often based on superstition and cultural inertia are still pervasive. The result is the perpetuation of inefficient modes of production as modernization is thwarted. The strength and influence of traditionally held beliefs is not surprising given that the acquisition of knowledge about agriculture by small-scale farmers is largely informal and oral (Edwards, 1961). That said, farmers have much more valid knowledge than they are typically given credit for. Their attitude toward innovation occurs on their own terms and the views of innovators or diffusion agencies and agricultural planners are overshadowed by their own assessments of the merits of a new idea or technique. Barriers between scientists and farmers need to be overcome for modernization to occur. These barriers can be removed through a two-way learning process and by increased respect for the ability of farmers to choose, improvise, and adapt. There also needs to be an acknowledgment that farmers have a significant complementary role to play in agricultural research, which goes beyond being passive subjects in studies.

Small-scale farmers in the tropics have often been stereotyped as being resistant to agricultural change and innovation. Very often the complex considerations that go into a decision to adopt or not are unclear to scholars and writers, who then attribute non-adoption to an inherent unwillingness to accept change by farmers. It often appears that the possibility of non-adoption has not been considered. Obviously, non-adoption or limited adoption is a distinct possibility, which should be considered. Farmers usually have more than one option and will chose an alternative that satisfies their needs to some degree. Also, it can take a while for an idea to catch on and some innovations are perceived to have more utility than others and will be adopted more quickly. Some will never be adopted as their comparative advantage to farmers is unsatisfactory.

It should not be a surprise that there was a close relationship between farmers' perceptions and inadequate information with many farmers' views being based on vicarious rather than personal experience. In the absence of strategic process of dissemination based on provision of accurate information, misinformation prevails and negative perceptions of innovations flourish.

Finally, the way rejection of an innovation is assessed needs to be re-evaluated. At what stage should farmers' non-adoption of an idea be classified as rejection? As we discussed earlier, some innovations take longer to catch on than others. The minisett technology was introduced to Jamaica in 1985, but to date, very little progress has been made in significant adoption rates among commercial yam growers. This is not surprising given the inadequate dissemination efforts. The views of farmers in our most recent study of the issue suggest that it is not inconceivable that a modified and well-marketed minisett package could become widely adopted by local yam farmers under the right agroconditions. Therefore, it might be the case that innovations rather than being rejected have simply not yet been accepted.

CHAPTER 6

Marketing and Distribution of Domestic Food Crops in the Caribbean: Implications for Food Security

6.1 Overview

Marketing and distribution of domestic food crops in the Caribbean requires urgent attention in the region's agriculture and food security debate and policy. The internal marketing and distribution is done through various informal commercial activities. The primary strategy is where farmers sell their produce to people, mainly women who sell in fresh produce markets across the country. Typically these women, known as *higglers* in Jamaica and *hucksters* in the eastern Caribbean, would go around to various farms and purchase different kinds of farm produce from farmers. Several of these higglers would then hire a truck/lorry, which would transport them to a designated market place. Some farmers now take their produce to these markets themselves, where the produce is sold in bulk to higglers or retailed to consumers. Also, an increasing number of farm families are setting up stalls on the side of major roadways, where they sell produce from their farms. Farmers who own pick-up trucks often load them with produce and park them in heavy traffic areas where consumers come to buy the produce. Other traders also purchase food from farmers for resale through this medium. The marketing and distribution of domestic foods have been identified as a major obstacle to production in the Caribbean. There is no regulated system in place, and small-scale farmers are basically left to their own devices in marketing farm produce locally.

One area of concern that is in need of special attention is the link between the tourism industries of the CARICOM countries and their agricultural sectors, especially the small-scale food-producing sector. Local tourism is booming while local agriculture stagnates and declines (Thomas-Hope and Jardine-Comrie, 2007; Dodman and Rhiney, 2008). This is a long-standing contradiction as evidenced by research in the 1970s and 1980s (see Momsen, 1972; Belisle, 1983, 1984). There has been some suggestions that in the past 20 years or so, there has been more of a willingness on the part of visitors to eat local cuisine (Momsen, 1998; Torres, 2003; Conway, 2004; Rhiney, 2009), but recent research indicates that we have a long way to go in getting local foods into hotel kitchens in a significant way (Rhiney, 2009). Farmers have argued that there is a need for a more proactive state role in the marketing and distribution of domestic food crops to enhance viability by providing stable markets and fair prices (Rhiney, 2009; Campbell, 2011). Government support should also encourage and educate farmers about establishing farming cooperatives for the production and marketing of food (Rhiney, 2009).

The state of marketing and distribution of domestic food crops is reflective of a general bias against the small farming sector. Traditional exports like that of sugar and coffee have established marketing boards, which buy produce from farmers in an arrangement that provides a steady income year round for farmers.

This chapter discusses the dynamics of food distribution and marketing in the Caribbean. We discuss how the current distribution and marketing system leads to declining production; we also suggest strategies for improving this vital aspect of agriculture in the region.

6.2 Factors Affecting Distribution and Marketing of Fresh Produce

Problems with marketing and distribution are linked to underdeveloped technologies and know-how in areas such as postharvest management, including storage facilities, grading, and packaging of produce.

6.2.1 Postharvest Management and Storage

De Lucia and Assennato (1994) define postharvest losses (PHLs) as quantitative and qualitative loss of food in the postharvest chain from the time of harvest through crop processing, marketing and food preparation, and even the final decision of consumers about what to eat and what to discard. The losses could be economic from the point of view of the farmer and nutritional in terms of consumption. PHLs may occur at any point in the postharvest chain and could be the result of several factors including improper handling or biodeterioration by rodents, insects, birds, and other microorganisms (Hodges et al., 2010). PHLs are not unique to the developing world, but in developed countries, discarding large amounts of food—that is, food left on the plate after a meal—is a big issue along with the problem of expired food (Hodges et al., 2010). Also in the developed world, PHLs are increasingly the result of peoples' awareness about and intolerance for substandard foods. In contrast, in the Caribbean, like in the rest of the developing world, the major issue seems to be inefficient postharvest practices resulting in loss of food, which would otherwise be eaten or sold, and loss of income to farm families (FAO, 2012).

There is no real data on the extent of PHLs in the Caribbean, but conventional wisdom holds that these are substantial. Unplanned production and the fact that many farmers grow the same crops at the same time is a major factor in PHLs in the Caribbean. Seasonal gluts are common and large quantities of produce taken to local markets remain unsold. Some of this will be eaten at home but often is discarded and fed to livestock. Fresh produce such as vegetables, fruits, and roots and tubers are naturally perishable, and considerable PHL is often experienced in the marketing and distribution system (Harris, 1988). Small-scale farming in the Caribbean is characterized by a lack of investment in storage facilities, and PHL of produce is a major problem facing local agriculture (Beckford, 2012; Rhiney, 2010; Harris, 1988). The lack of storage facilities makes it impossible for farmers to properly coordinate their production schedule, leading to a cycle of oversupply and market gluts and scarcity and high prices of some crops at different times of the year. This lack of coordination is a significant reason for the poor linkages between local agriculture and tourism as farmers are unable to provide consistent and reliable supply of food to hotels and restaurants that cater to the tourist market (Rhiney, 2009).

Poor transportation adversely impacts food marketing and distribution. The lack of refrigerated transport vehicles causes spoilage of fruits and vegetables. In the eastern Caribbean, PHLs could be considerably reduced by the introduction of modern refrigerated freight liners and the use of appropriate packaging (Harris, 1988). Typically roots, tubers, and grains are less perishable because they have less moisture content, but poor postharvest handling may lead to losses in both quality and weight (Hodges et al., 2010). For example, peas and beans, which are stored for long periods in poor condition, grow mold and are infested by weevil. Loss of tubers and roots is also caused when produce has to be cut to prepare for sale and to satisfy the consumer. For example, in Jamaica, where yam is an important export crop, export agents do a lot of paring and carving, which result in unquantified amounts of loss of edible produce.

Reduced PHLs can lead to increased food availability by reducing physical losses and increasing income (World Bank, 2010). Instead of producing more, reducing PHLs could save productive resources and reduce environmental stress (Hodges et al., 2010; Stuart, 2009; Buzby et al., 2009; Boxall, 2002). While there is no accurate picture of the extent of PHLs in the Caribbean, according to the International Rice Research Institute (IRRI), PHLs in Asia, Latin America and the Caribbean, and Africa reach up to 30 percent and are caused by losses to animals and pests and contamination. In Jamaica small-scale experiments with farmers showed that yams experience loss of weight when kept for too long. A box of yams weighed 25.6 kg on Sunday and 24.8 kg on Saturday when stored indoors and 23.8 kg when stored in the sun. Local yam farmers try to mitigate weight loss by covering yam with grass to protect it from the sun (see figure 6.1).

Farmers explained that over time, the yam "dries out" or loses moisture and the weight drops. Weight loss varied according to the size of yams, with smaller yams losing weight at a faster rate than large yams. This is supported by Harris (1988), who posits that water is lost from all crops after harvest and the process starts immediately after harvest. Wilting and shriveling of produce is a sign of water loss and breakdown. Water loss is accompanied by breakdown of the food material, which eventually leads to decay (Harris,

Figure 6.1　A farmer covers fresh produce to reduce postharvest weight loss

1988). Temperature is a major factor in water loss and decay, and therefore, tropical agriculture has naturally high rates of water and material loss and consequently high rates of PHLs. The lack of proper storage and packaging and transportation facilities exacerbate the problem. Simple techniques like harvesting during the cool of day, keeping produce as cool as possible, and getting produce to the market as soon as possible can help to mitigate PHL.

Postharvest systems in the Caribbean require urgent attention. Educating farmers about the causes of PHLs and postharvest handling of and care for food is also critical. Research into the extent of PHLs and the impact on food security and income is needed. Investment in postharvest storage facilities is crucial. More formal markets and formalized marketing arrangements are needed to significantly reduce PHLs. There is a great incentive to reduce PHLs as there is a direct link between such losses and rural livelihoods, food security, including availability, safety, and nutrition. On-farm experimentations in PHL reduction strategies are necessary and should account for the effect of climate change.

6.2.2 Grading and Packaging

Small-scale food marketing in the Caribbean is characterized by an absence of sorting and grading of fresh produce. This is the case at all stages of the distribution chain—farmer to higgler and higgler to consumer. Education and training in the proper cleaning, sorting, and grading of fresh produce is therefore needed. This would allow farmers to market produce differently based on quality. Currently there is no sorting or selling according to grade, which implies that all produce is sold for the same price regardless of the quality. Typically, farmers deliberately mix produce so that there is a mixture of different quality in every unit sold. Packing is an important factor but is neglected in the Caribbean. Packing should facilitate easy handling of produce and provide protection from damage. Most farms are too small to even have packing houses, and packing standards, methods, and containers are rudimentary at best. Produce are often battered and bruised by the time they get to the market.

6.3 Systems for Marketing and Distribution of Domestic Food in the Caribbean

6.3.1 Higglering/Huckstering

Higglering is the main system employed in the Caribbean for the internal marketing and distribution of domestic foods. The system has its origins in the food markets where slaves sold produce from their provision grounds and thus might have African origins (Besson, 2003; Mintz, 1989). Slave owners provided land for their slaves to grow some of their own food. Although largely a subsistence practice, some slaves grew enough to barter in local market places. It became customary for male slaves to cultivate food and for female slaves to sell or trade this food on Sundays, which is the only day slaves did not have to work on the sugar plantations. This marketing system expanded greatly in the period immediately after emancipation, when ex-slaves established free villages and started the development of an independent peasantry. This system of marketing in the contemporary Caribbean reflects both continuity and change from the slave era (Besson, 2003).

A higgler may be described as someone engaged in the purchasing and resale of goods as an informal merchant. In terms of food marketing, higglers buy wholesale food directly at the farm gate and resell it on a retail basis in local food produce markets. Most higglers are women, and the term is used to describe female marketers of agricultural produce, but at any given point many men will be seen selling food in produce markets. These men are mainly farmers selling the produce from their own farms. Women, on the other hand, typically market food from the farms of their spouses and resell food they have bought from other farmers.

Higglering is an internal marketing system dominated by women in their role as marketers in the local food production system. This role is essential, though perhaps trivialized and underestimated in rural development planning (Gonzalez, 1985). It is a visible manifestation of female participation in economic production in the Anglophone Caribbean (Gonzalez, 1985). A study of higglering brings to light female domination of the distribution trade for fresh farm produce in the region. Because of this, higglers are often referred to as "market women" in local parlance.

Many higglers develop a loyal clientele over time. They know which days these customers shop and what products they want in what quantities. This relationship is cultivated over time and higglers go to great lengths to protect it. Conflicts between higglers are not uncommon if one perceives that another is moving in on one of her regular customers. Some customers will buy only from certain higglers who try their best to ensure that they have what their customers need so that they do not have to consider buying from others. A regular or "good" customer can always count on "brata" from the higgler. Brata is extra—if you buy two dozen oranges, the higgler will throw in a few free ones. This is an important part of keeping a regular customer. Higglers will go to market even when sick so as not to disappoint their good customers (Gonzalez, 1985). When certain produce is scarce, they will make special efforts to secure some quantity for their good customers. These will be tucked away somewhere out of sight until the customers show up.

Higglering is a loosely organized and informal system for marketing and distributing agricultural produce domestically. Fridays and Saturdays used to be the market days, but markets are now opened everyday, except on Sundays, with modest activity from Monday to Wednesday, increased activity on Thursday, intensifying significantly on Friday and Saturday. In the major urban centers, there is considerable activity at markets throughout the week. Higglers from rural areas arrive on Thursday or Wednesday and depart Saturday evening. Many higglers are local women and men who come to the market place and buy wholesale from farmers or travel to the country to buy food and take back to the urban market. Produce from urban gardens is becoming more widespread in food produce markets.

A rural higgler typically goes around to several farms in and around her community and arranges with farmers to purchase stock to make up a "load" for the weekend market. She makes arrangements for certain quantities of specific foods and negotiates prices. Goods are taken on credit with payment made on her return from market. A market trip could last from Wednesday or Thursday to Saturday, but higglering is a week-long activity. The higgler spends Monday to Wednesday securing a load and then the next three days at the market sleeping on rudimentary bedding on the floor near her stall.

Higglers travel to the market houses in large trucks or lorries, with their produce stacked around them. Typically a truck is hired by several different higglers. The truck goes around and pick up each higgler and her produce. Trucks are often overloaded and produce is packed in sacks or boxes. Damage to produce is common. Trucks frequently break down, and a simple punctured tire can spell disaster as under the heavy load, a routine activity like changing a flat tire becomes a herculean task. If the problem is more severe, the truck and its cargo might not even make it to the market. In such situations, everyone loses. The higgler is stuck with fresh produce, much of which typically is quickly perishable. With any luck, some produce may be able to make it to the next market trip a week later, but this is rare.

The higglering system is unpopular among farmers. Payment is not guaranteed because much depends on the fortunes of the higgler. Sometimes farmers are not paid as higglers often claim to have experienced "bad market" caused by inclement weather or market gluts for some produce. Many produce markets are not in the best physical shape. Some have leaky roofs and unpaved underfoot and so rainfall leads to muddy slippery conditions, running water and puddles, and misery all around. Heavy rains ruin market days as consumers stay home and sales drop.

Produce markets are operated by local municipalities and are found in major urban centers and town, but also in smaller rural towns. These are designated market houses that sellers pay to enter and typically occupy a particular spot week after week where they set up a stall. This allows their customers to locate them easily. In the Caribbean, like in many places in Africa, the first markets were usually associated with busy road junctions (Gonzalez, 1985).

6.3.2 Farmers' Direct Wholesaling

Direct wholesaling has become increasingly popular in countries like Jamaica as a response to dissatisfaction with the higglering system. In this system of marketing, farmers take their produce to the market places, where they sell directly to higglers on a wholesale basis. In many cases, produce has been pre-ordered and exchanges are quick. The higglers involved in this trade are usually urban people, but rural sellers who had traveled on trucks and got rid of their stock quicker than anticipated often top up their stocks of certain items based on the trends in demand that they have observed. Some rural higglers also prefer this system because they do not have to worry about securing a load and hiring a truck. They are able to go to the market place sometimes without any produce or with a light load and buy wholesale from farmers. This is a more efficient system for them to save time and money, and to increase their income.

This system of marketing allows farmers to make more than one trip in a few days. For example, a farmer who takes produce to the market on Wednesday evening or Thursday morning may gauge his ability to sell more produce, and depending on the distance to his farm, he might go home and return on Friday morning with more produce. Farmers who do this are often responding to orders from higglers. The farmers' wholesale system cuts out the intermediaries and increases profits and income to the farmer. It also brings the farmers in closer contact with the market and gives him a better sense of market dynamics like demand and supply for certain crops. Farmers and consumers interact and communicate, and this brings obvious benefits to both. Farmers are also able to see trends in prices and are in a better position to bargain with higglers. Food should become more affordable to the consumer with the intermediary gone from the supply chain.

6.3.3 Farmers Markets

Farmers' markets have been popular in developed countries for years but have not enjoyed the same popularity in developing countries. In the United States, for example, there was a 9.6 percent growth in the number of local farmers' markets between August 2011 and August 2012 (US Department of Agriculture, 2006). The top states included California, with 827 farmers' markets and New York with 647. Farmers' markets benefit consumers, farmers, and communities (Friends of the Earth, 2011; Robinson and Hartenfeld, 2007). Shoppers can procure locally grown, top-quality farm fresh produce at more reasonable prices directly from the growers. Farmers get to connect with consumers and raise their income. Communities benefit as these markets bring people together and supply local food (Halweil and Thomas, 2002). In the Anglophone Caribbean, farmers' markets are now just coming to the fore even though they have been established in nearby Cuba for many years. In Jamaica, two farmers' markets have been established in the past three years. They have proven to be very popular among both farmers and consumers and more are being planned. Farmers' markets could be a game changer in the internal food marketing network in the Caribbean. These markets provide an avenue for farmers to interact with and sell directly to the consumer. They cut out middlemen and third-party traders, creating a distribution system that benefits both the farmer and the consumer. The produce available in farmers' markets have a reputation for being farm fresh and local, which is a major draw for buyers.

6.3.4 Fruit and Food Stands

Fruit stands are popular features of Caribbean rural and urban landscapes. Although difficult to quantify, this is an increasingly popular way to market fresh farm produce. These stands are different from the stands that are located on urban sidewalks and pavements in large cities in North America and Europe and include fruits, vegetables, and horticultural products. The Caribbean versions are roadside stands on high-traffic public roadways, street corners and busy intersections, and bus stands. They sell a wide variety of fruits and vegetables but also a variety of staple foods including roots and tubers. The kinds of fruits and the variety of products found depend to a large extent on the part of a country you are traveling through and the time of year or the season. Fresh produce typically come from kitchen gardens, fruit orchards, and farms of the sellers. Some stands are mainly fruit stands, while others will have both fruits and vegetables and still others would have fruits, vegetables, and "food" such as a variety of roots and tubers. Here too the proprietors tend to be overwhelmingly women.

This form of marketing is an informal self-help approach that serves farmers well. Fresh produce stands are often located at the front of homes, giving women the flexibility to combine this activity with other household roles. It is common to find young children tending to the stands while their mothers are otherwise engaged. Sometimes a cluster of stalls operated by different individuals have been set up in very close proximity to each other, creating a kind of fruit stall arcade. These stands are dependent on traffic. As is the case with higglers, fruit stall operators cultivate special relationships with regular motorists who will buy only from one or a select few. The relationship among sellers is interesting. They are in obvious competition but will provide change for large bills to each other and will even loan a competitor produce to sell to a buyer. In recent times popular fruit stands have been closed because of health and sanitation concerns. We would argue that the importance of these fresh produce stands in the supply chain is not fully understood and appreciated and that this is an area requiring more research.

6.3.5 Supermarkets, Hotels, and Restaurants

These types of marketing are not significant parts of the food supply chain in the Caribbean. The linkages between tourism and local agriculture are very weak and have very little mutual stimulation. Hotels and upscale restaurants use mainly imported food, and fresh produce in local supermarkets is still a fledgling concept as most Caribbean folk prefer to get their tropical fruits, vegetables, and traditional foods from local produce markets. The foods available in the fresh produce markets are fresher and far less expensive than those available in supermarkets. Still in major urban centers around the region, more local farm produce are making their way onto supermarket shelves. Where farmers have been able to establish production and marketing groups like in parts of Jamaica and St. Lucia (Rhiney, 2009; Timms, 2006), there has been some penetration of the tourist market, but hotels, restaurants, and supermarkets do not play a significant role in the supply chain of locally produced food. The reasons for this are discussed at length in Chapter 16, but speak to the urgent need for regional agricultural planners to deliberately plan for intersectoral linkages and treat food marketing and distribution as an agricultural activity with implications for food security and rural livelihoods.

6.3.6 Exporters

In some islands of the region, some domestic food crops have international markets. For example, in Trinidad and Tobago and Guyana, rice is grown by small-scale farmers and is marketed both internally and in the export trade. In the eastern Caribbean—St. Lucia, St. Vincent, and Dominica—and in Jamaica, bananas have an internal and an export market as well. Many root crops and tubers and vegetables in Jamaica that were traditionally grown for domestic consumption are now part of a category of agricultural crops known as nontraditional export crops. These include yam, dasheen, sweet potato, and pumpkin. Yam is the major nontraditional export crop in Jamaica earning over US$15million in 2009. The main subspecies of yam cultivated is yellow yam (*Dioscorea cayensis*), and Jamaica has a monopoly over the world market for this species. Other major yam exporters such as Nigeria, Ghana, Brazil, and Costa Rica produce various varieties of white yam. Yet only around five percent of the total national production is exported, which emphasizes the tremendous potential to develop more export markets. Most yams are exported to the United States, Canada, and England, which have large Caribbean and African diaspora populations.

As is the case with the local market, the export market is generally informal and loosely regulated. In the case of Jamaican yams, export agents or independent operators purchase yams from farmers and sell to export packaging plants. Unlike the case with higglers, this is a cash-on-delivery business, which provides greater guarantees for farmers. Some farmers sell directly to packaging plants, thus cutting out the middlemen and increasing their income. These farmers are sometimes employed by the packaging plants to procure supplies.

6.4 Summary

The whole internal food marketing and distribution system in the Caribbean needs to be examined. The link between food marketing and distribution and food production should not be underestimated. Studies have repeatedly identified a lack of adequate food distribution infrastructure as a major disincentive to increasing agricultural production. Without adequate market access, agriculture is not a viable economic activity. The internal market structure for domestic foods is informal and unregulated, and reflects farmers' and local peoples' self-help strategies. Municipalities typically provide the market houses, but too often that is the extent of any state involvement in the internal food distribution network. The higglering system is accepted for what it is and serves a vital function. Right now it is the main avenue for internal food distribution. Without it small-scale farming would crumble. Older Jamaican farmers still lament the closing of the Agricultural Marketing Cooperation (AMC) in the 1980s. The AMC was a state-run marketing board that bought fresh produce from farmers at respectable prices and represented an alternative to the higglering system.

Caribbean governments need to do much more to improve internal marketing systems. Exports such as coffee, cocoa, coconut, citrus, sugar, banana, nutmeg, pimento, and so on, all have commodity boards. Nothing like this exists for traditional domestic food crops. Indeed a farmer could have different experiences marketing the same crop depending on whether he/she is dealing with the export market or the domestic market. For example, in the eastern Caribbean the production, packaging, assembly, transportation, and distribution of bananas for the export market are well established through the Windward Island banana growers' board WINBAN and the Geest Organization (Harris, 1988). But the sale of bananas locally in the various islands is informal and unstructured.

Governments should also do more to secure international markets and take advantage of the large and growing Caribbean population in Europe and North America. Promotion of inter-regional trade in food is critical and should be a key element of the regional food security strategy. Traditional foods like yams, dasheen, banana, plantain, coco yam, and pumpkin are very popular among the Caribbean diaspora in the metropolises of North America and Europe. Agroprocessing should be encouraged to add value and raise rural and urban farm incomes. Postharvest food management and storage also need to be addressed. Too much food is wasted due to spoilage and there are no mechanisms among small-scale food farmers to safely and effectively store food for strategic release to the market as conditions dictate or in response to emergencies. The presence of secure markets for their produce is a strong incentive for local farmers to produce more and enhance food availability. Weak market infrastructure is a fundamental factor in food production decline and also explains the poor linkages between local agriculture and the world-famous Caribbean tourism industry.

Infrastructure is also necessary—market houses need to be improved. Better policies and infrastructure to connect farmers to markets are required. Appropriate marketing institutions must be developed. Farmers' marketing groups and collectives should be encouraged and promoted to allow farmers to better respond to market conditions. More research is needed into the role that farmers' marketing groups and cooperatives can play in more efficient food production and marketing. According to Rhiney (2010), these groups once known as Production and Marketing Organizations numbered 800 in Jamaica in 2008 according to the Ministry of Agriculture data and had more than 30,000 farmers. In St. Lucia, research has also shown successful farmers' cooperatives (Timms, 2006). The way they function and operate and the extent to which they have been effective in marketing food is something that requires more research, but indications are that they have achieved varying success in, for example, breaking into the local tourism market (Timms, 2006; Rhiney, 2009).

The advent of farmers' markets could play an important role in food marketing and distribution. Cutting out the intermediary eradicates many problems associated with the higgler marketing system. It facilitates direct contact between farmers and consumers, leading to fairer prices for the consumer and greater income for the farmer. Higher incomes will make farming more profitable and more attractive and influence farmers to increase production. A major problem with food marketing in the Caribbean is lack of communication between farmers and consumers and lack of knowledge of market conditions and consumer preferences and needs. Farmers' markets facilitate better communication and a better understanding of each other's needs.

CHAPTER 7

Women, Agriculture, and Food Security in the Caribbean

7.1 Overview

Sustainable food security in the Caribbean requires the effective participation of women in food production. This is significant in the context of the dimensions of availability, access, and supply of nutritious foods and the implications for the overall household food security. There are many commercial female farmers in the Caribbean, but women are mainly involved in the marketing and distribution of food as they make up a disproportionate amount of sellers in the local produce markets across the region. The strategic participation of women in food production could be effective for addressing food security and nutrition at the household level. The aim would be to increase supplies of nutritious foods for their households. Women should also receive training in food handling and preparation to maximize the nutritional value of their families' meals. A recent project by the International Development Research Center (IDRC) in Lebanon speaks to the importance of women in food security and nutrition (Boothroyd, 2010).

An interesting aspect of the IDRC study, which holds lessons for the Caribbean, is the development of a communal "Healthy Kitchen" by women in three villages. The project centered on the preparation of traditional dishes using wild edible plants and other produce. The women obtained training in commercial food preparation and marketing and became nutritional ambassadors, selling their produce in local markets, catering at weddings and other events, and operating an eco-lodge, which showcased their Healthy Kitchen cuisine. This is the kind of approach that may be necessary to bring Caribbean populations back to local and traditional foods. It is different from conventional eat local campaigns conducted through the media, which have been largely unsuccessful due to the top–down approach and a lack of meaningful grassroots community and household engagement (Beckford, 2012).

Most of the food produced in the Caribbean is sold as fresh produce. An important component of food security and women's participation should be initiatives to promote increases in value adding agroprocessing at the household or community levels. Agroprocessing would drastically reduce postharvest losses, preserve food, and add value, thus increasing farm incomes. Cottage industries based on locally produced fresh farm produce should be promoted, encouraged, and supported. In this regard there should be efforts to establish properly constituted cooperatives, but household-level industries should also be pursued. Women could also play an instrumental role here as the history of cottage industries in the region suggests that they have always taken a leadership role. Again there will be a need for training and ongoing learning in areas like food processing, business management, marketing and distribution, and accounting.

This chapter examines the traditional and changing roles of women in agriculture in the Caribbean. It explores the role women play and can play particularly in enhancing household and community food security and nutrition. We make the case that sustainable food security and nutrition in the region can be significantly enhanced by the increased and strategic participation of women in food production and village and household-level agroprocessing and cottage industries. To be effective and sustainable, food security strategies in the Caribbean must create the conditions for females to improve their own food security and that of their families, improve nutrition, and achieve greater economic independence. Much has been made of women's contribution to household nutrition. We argue that while this is critical, women can play a bigger role in the overall food security and contribute more to improvements in household income and rural development.

7.2 Role of Women in Agriculture: A Global View

While the overall proportion of the labor force working in agriculture has declined worldwide, the proportion of women is actually growing as they face restrictions in other areas of the economy. Half of all farmers in the developing world are women, and women farmers could produce as much as 30 percent more food if they had access to the same resources as men (Gates Foundation, 2012). This would lead to improved agricultural production by 2½ to 4 percent and would reduce undernourishment by 100 to 150 million people in the developing world (FAO, 2011). Veveer (2011) also suggests that food productivity could significantly increase if women were given more resources. Furthermore, research indicates that women are more likely than men to use their incomes to improve the well-being of their families (Center for Universal Education at Brookings, 2011; Veveer, 2011; Kumar and Nair, 2004).

Agriculture is central to economic growth when women can learn the best way to produce their own nutritious food. Closing the gender gap in agriculture is fundamental to increasing food availability and enhancing food security. Women shoulder a significant portion of the agricultural endeavor in developing countries and should receive more support to make them more productive (Veveer, 2011). Reduction in undernourishment requires the participation of women in agriculture on an equitable basis. This is significant given the important role women play in agriculture and the economic constraints they face.

The recognition of the important role of women in agriculture is manifested in the development of the Women's Empowerment in Agriculture Index by Feed the Future in collaboration with the International Food Policy Research Institute, the Oxford Poverty and Human Development Initiative, and USAID. The index was piloted in three countries on three continents— Uganda in Africa, Guatemala in Central America, and Bangladesh in Asia— and measures change in five domains—women's household decision-making, access to resources including credit and land, adequacy of women's income in relation to food access, community leadership roles, and women's labor time allocations (Veveer, 2011).

Women constitute on average 43 percent of the agricultural labor in the developing world but face daily constraints, which have deleterious effects on their ability to produce food. These constraints include restrictions to their ability to buy, sell, or inherit land; save money in financial institutions; access loans; sell crops; obtain fertilizers and other agricultural inputs and benefit from extension services. The result is that yields from women's farms average about 25 percent lower than men's farms.

It should be noted that increased output of food is necessary, but not sufficient to combat food insecurity. There have been significant increases in food production in some parts of the developing world, but this has not led to concomitant increases in nutritional status. According to the World Bank, the ignorance of the value of women as producers and providers of household food supply is a major reason for modest gains in nutrition in South Asia, while food production has increased significantly. The same scenario also plays out in sub-Saharan Africa. In Kenya, for example, efforts to boost milk production by introducing hybrid cows led to male domination of dairy farming, and while milk output increased, nutrition levels declined. The explanation for this is that when women controlled dairy production, they dedicated the morning milking to the needs of the household, especially children, while selling the evening milk (Global Conference on Women in Agriculture, 2012).

International development agencies perpetuate the problem sometimes unwittingly by assuming that farmers are men and leaving gender considerations out of their research and development packages (Global Conference on Women in Agriculture, 2012). Ruth Mienzen-Dick, senior research fellow at IFPRI, has called for a change in the thinking of donors, researchers, and aid groups to recognize women as farmers and to see their activities as critical to food security. For example, in many developing countries, women are the primary actors in postharvest activities, but these are not high on the agricultural research agenda, which is ironic given the colossal losses of food in developing countries caused by a lack of postharvest expertise and facilities. Postharvest losses of grain in sub-Saharan Africa totaled some US$4 billion between 2005 and 2007, which was about the same as the value of cereal imports to the region over the same period (Global Conference on Women in Agriculture, 2012).

Women play a significant role in agriculture all over the world as producers, agricultural workers, marketers, and food processors. Despite this, their roles in these economically important activities have tended to be downplayed and obscured. Women have traditionally been marginalized players in the area of decision-making and policy-making, resulting in a devaluation of their role in all aspects of food and agriculture.

Rapid modernization of agriculture through the introduction of new technology benefits rich farmers more than poor ones and men more than women. This is especially the case where the new technologies lead to greater commercialization of farming as this shifts agricultural income and profits from women to men. This leads to adverse consequences for rural farm families as women's income benefit their families more than men's income. In the global context, women's role in agriculture goes well beyond cultivation. In other words, women in agriculture are not just farmers. They play important roles in all farm-related activities from land preparation to marketing and their activities often add value to farming. They contribute a higher proportion of labor in agricultural sector than men but their participation in decision-making is limited (Satyavthi, Bhardwaj, and Brahmanand, 2010).

7.3 Women and Agriculture in the Caribbean

In the past 45 years there has been a great deal of research on Caribbean agriculture. An examination of the work done over the years reveals that there is a lack of research about the role of women in Caribbean agriculture. More interestingly, the more recent work done in the past 15 years, for example, rarely has any significant consideration of gender issues in agriculture and food security in the region. The existing research on women agriculturalist in the region is therefore quite dated, but relevant, as it gives some historical context and can provide insights into topics in need of research.

According to Momsen (1995) one-third of small-scale farmers in the Caribbean are women who grow food mainly for home consumption and local sale. Momsen suggested that women typically cultivate the most marginal lands—small, infertile, and inaccessible. Momsen posited then that whereas governments in most regions of the developing world were working to reduce the barriers to women's participation in agriculture, the reverse seemed to be the case in the Caribbean, where citing her research in Barbados, she argued that there were traditional gender inequalities in land ownership and agricultural production. Barrow (1994, cited in Momsen, 1995) claimed that women farmers were experiencing marginalization through restricted access to technology and scare resources and their participation in agriculture was being impeded by ideologies of female domestication.

In research from Belize, Bliss (1992) pointed out that Black Carib women—known as Garifuna—have maintained a stronger tradition in farming than men for generations. These women have traditionally operated subsistence plots planting a variety of staple foods that have been important in their food security. This is supported by earlier research that showed that it was traditional in Garifuna culture for women to be the primary food producers with the responsibility for land preparation, planting, and harvesting (Gonzalez, 1969). This research also establishes the traditionally subsistence orientation of these farming activities. In the Garifuna culture, women got very little help from men as many migrated in search of work. Where men were involved in agricultural activities, they provided protein through fishing and engaged more in commercial farming enterprises (Bliss, 1992; Gonzalez, 1969). The story of women farmers in Barbados is also punctuated by evidence of marginalization. Momsen in an analysis of data for the period 1847–1916 found that the farms operated by women were generally smaller than men's farms and were situated on the most marginal farmland (Momsen, 1995).

In Belize about 80 percent of women farmers work their land jointly with their spouses, but it was found that there were a few tasks—typically between two and four—which they were responsible for (Bliss, 1992). Women had their own farms only after they married or were cohabiting with a man who would provide a piece of his holdings for her (Bliss, 1992). Typically, women depend on men for land clearing and preparation in Belize, which is a feature of Caribbean small-farming systems and somewhat different from other areas in the developing world such as Asia and sub-Saharan Africa. This meant, though, that the size of a woman's farming operation was dependent upon the amount of land men were willing or able to prepare for her. Bliss (1992) reported that there was a clear gender division of labor in terms of crops planted. Some crops were clearly women's crops, and others were men's crops, although this did not mean that they were exclusive to either women or men. For example, root crops were more often grown by women while rice and corn were predominantly the domain of men. Women made the decisions about the types of crops they wanted to plant and typically had their own plots that may have been located on or off their partner's land. These plots were cultivated with women's crops like cassava, yampee, yam, and coco-yam, which were solely used for household consumption and thus played a vital role in household food security, including nutrition. Bliss (1992) explains that as women grew older, they planted a smaller variety of crops focusing on low-maintenance crops especially cassava, which once planted requires little attention.

In the Caribbean, data for the involvement of women and men in agriculture are generally confined to transactions in the formal market sector. Statistics related to the production and marketing of the principal export crops, such as banana and sugarcane, indicate that men are predominant in the control and marketing of export crops or cash crops. For example, a 2002 social audit in the sugar industry in St. Kitts/Nevis showed that men were the major source of labor by a considerable margin and the same situation existed for banana, the other major export crop (FAO, 2011). On the other hand, women were major participants in the cultivation of domestic food crops and subsistence cultivation, and were involved in the regional export trade in food buying from farmers directly and exporting to other regional territories such as Grenada, St. Vincent, Trinidad and Tobago, and Barbados (FAO, 2011).

As is the case in most developing countries, we would argue that despite the absence of recent research on the subject, there is enough evidence of the fact that women engaged in agriculture in the Caribbean play important roles in household food security as income earners, nutritional providers, and managers of natural resources and biodiversity, although the success with which they are able to execute these roles is too often hindered by restricted access to resources and services such as land, capital, technology, and extension service. For example, the FAO reports that in Jamaica the majority of women farmers—principally engaged in production for home consumption—are smallholders with the average farm size being significantly smaller in area than that held by men. This would suggest that production constraints related to land tenure and access clearly tend to impact more heavily on women than on men. Evidence points to the fact that while women are predominant in subsistence cultivation for home consumption and in the domestic sphere of agricultural marketing, men are typically more actively engaged in the marketing of traditional and nontraditional agricultural commodities to regional and international markets (FAO, 2011). That said, agricultural research in the Caribbean in the past decade has shown that there are many women who are bona fide commercial farmers and women's roles go beyond subsistence cultivation and food marketing and distribution. Figure 7.1 shows a female farmer at work.

According to ECLAC (2005), "female gender roles and functions in agriculture and rural development are generally aligned to non-remunerative, non-economic, unpaid work and occur within the informal sector of the economy" (p.14). ECLAC suggested that women producers in the Caribbean experience differentiated access to essential resources and technology and might be especially impacted by trade liberalization in the region (2005). They argue that in the production and marketing of agricultural produce in the Caribbean, women actually engage in activities that are more diverse than men's as they are involved in a wider range of crops and seek more markets than men (ECLAC, 2005). Because women are almost exclusively responsible

Figure 7.1 Female farmers at work

for social reproduction, they possessed the potential to take advantage of emerging market opportunities (ECLAC, 2005, p. 40).

Portilla and Zuniga (2007) argue that the recognition of women's role in agriculture did not really develop until the 1970s, which was declared the Decade of Women by the United Nations. Campillo and Kleysen (1996) in a study of women food producers in Latin America and the Caribbean concluded that the contribution made by women farmers is still largely invisible. The study found that many women operating small-scale family-oriented plots were not counted in the economically active population (EAP) and that their contribution to food production was underestimated in human power assessments. Women's work on family food production plots was treated as an extension of their domestic duties, and because they are not paid a salary, their contributions were not accounted for in official counts of economic activities (Campillo and Kleysen, 1996). As is usually the case when the terms *Latin America* and *the Caribbean* occur in the literature, data from this study was dominated by Central and South American countries but did include Jamaica and Trinidad and Tobago.

Efforts to recognize and better support women food producers in the Caribbean have led to the formation of women farmers' groups in the region. For example, the Trinidad and Tobago Network of Rural Women Producers (NRWP) has been supported by the Inter-American Institute for Cooperation in Agriculture (IICA) since 2000. The NRWP is a member of the Caribbean Network of Rural Women Producers (CANROP), an umbrella organization for national chapters of rural women's association in the Caribbean. CANROP was launched in 1999 in response to the need to equip women with education and support to enhance their socioeconomic status and the livelihoods of rural farm women (www.agricarib.org/caribbean-network-rural-women-producers-canrop). In addition to Trinidad and Tobago, it has affiliates in Barbados, Grenada, Guyana, Jamaica, St Lucia, St Vincent and the Grenadines, and Suriname. CANROP seeks to empower rural women to improve their standard of living through training, access to small loans and credit, networking, and cultural exchange (www.agricarib.org/caribbean-network-rural-women-producers-canrop). International organizations like the FAO and IICA have been active in recent times in promoting a more gendered policy response to food and agriculture and food security in the Caribbean.

The status of women in agriculture in the Caribbean is still not fully understood because there has been very little recent research in this area. This is true of the food and agricultural practitioners as well as academics. The lack of gender-differentiated data makes it very difficult to assess how gender may be at play in poverty and food insecurity in the Caribbean. In the contemporary Caribbean, the role of women in agriculture is most often defined in the marketing and distribution of agricultural produce. Domestic foods are marketed mainly in produce markets located in major towns in both rural and urban spaces. These produce markets have their origins in the market system established first during slavery when slaves marketed the surplus from their provision grounds and then the markets further developed after emancipation, with the establishment of the free peasantry in the English colonies around the region.

According to Besson (2003) there is some controversy about African heritage and gender differential in the domestic food marketing system in the Caribbean. Besson argues that the predominance of women in the marketing and distribution of domestic food locally has been attributed to African cultural influences, but she cites Mintz (1980, 1989) and Mintz and Price (1992), who argued that although women dominate the food marketing scene in Jamaica and Haiti in the post-emancipation period in a complementary role to male cultivators, there is no evidence that women outnumbered men and the situation was more like an even split of the genders.

After emancipation there was an increasing tendency toward women marketing and men cultivating as division of labor in agriculture became more gendered (Besson, 2003). Mintz and Price's (1992) study of Haiti and Jamaica found that after slavery women were overwhelmingly dominant in food marketing while men were equally dominant in cultivation (Mintz, 1960, 1989, cited in Besson, 2003). Besson cited Mintz (1989) and Durant-Gonzalez (1983), who wrote that marketing and higglering provided women with opportunities for autonomy, mobility, and relief from domestic service.

The current dynamics of gender in agriculture can be established only through careful research. Intuitively, it is common perception that food marketing is dominated by women while men dominate cultivation. Unfortunately this question has not been the subject of recent research in the region. In our view the role of women in food marketing in the Caribbean is an essential component of food and agriculture, is more than just a complementary role, but is just as valuable as the cultivation of food itself. We argue that food production implies both the cultivation and marketing of food. The first component is growing the food. The second is making it available to the public. For many rural households and farm families, farming is a family enterprise. Men may very well be the dominant cultivators, but women are partners in providing light labor, nourishment, and sustenance for the male cultivator and form the last link in the production chain—marketing and distribution. In our own research from Jamaica, in which ethnographic research through participant observation has been a major methodological approach, we have worked in fields where children and women do light work, and women come out in the morning and leave at lunch time after preparing the mid-day meal for men.

The situation in the Caribbean is more complex than in other parts of the developing world. While it is clear that gender bias exists and is problematic, the status of women is perhaps not as marginalized as elsewhere in the developing world. For instance, Caribbean women have obtained access to land and have cultivated home gardens and provision grounds since the days of slavery (Besson, 2003, 1995, 1992; Momsen, 1988).

7.4 Increasing the Role of Women in Agriculture

"Ensuring gender equality in agriculture is not just the right thing to do it is also the smart thing to do. When the status of women improves, agricultural production increases, poverty decreases, and nutrition improves. Unleashing women's potential by closing the gender gap in the agricultural sector is a win-win strategy."

(Veveer, 2011)

7.4.1 Recognizing the Role of Women in Household Food Security and Nutrition

There is a need for urgent action in increasing the role of women in enhancing household food security. Evidence-based information on the value of women's contribution in this area and the expanded role they can play should be disseminated to make their contribution and potential more visible. Furthermore, this knowledge should be mobilized into action with the overall goal of increasing women's participation in food and agriculture, providing greater support for their participation, and consequently increasing their contribution to food security. The inaugural Global Conference on Women in Agriculture (GCWA) 2012 suggested that the issue of household food security and nutrition is of paramount importance in consideration of gendered agricultural policy (GCWA—Synthesis Report, 2012). We would suggest that the Caribbean approach should adopt this view and focus on creating a strategy for increasing the capacity of women to significantly influence their household food security in terms of both increasing food availability and enhancing nutrition. The role of women in the nutritional status of children should be given special consideration in this regard. There is a need for women to be aware of the roles they play and can play in the food security of their families and the nutrition of their children and themselves in all phases of their fertility cycle. Women have been invisible in agricultural leadership, research, extension, and decision-making, and this has in turn led to their exclusion from and neglect by agricultural policy and in the Caribbean academic research. The Caribbean has a food security strategy, but the role of women agriculturalists is not a component. There is no discussion of the contributions of women to food security and nutrition and no strategy for enhancing this role. This is a grave mistake.

7.4.2 Food Marketing and Distribution

Caribbean women play important roles in agriculture outside of food cultivation. None of these roles is more important than the marketing and distribution of domestic foods, which is dominated by women in the contemporary Caribbean (Besson, 2003; Brown-Glaud, 2011; Durant-Gonzalez, 1983; ECLAC, 2005). It is fair to say that in the Caribbean, small-scale food farming is very much a family affair and women play other roles in support of their male partners who typically do most of the cultivating. Women also engage in small-scale informal agroprocessing, but it is food marketing and distribution where they play their biggest role. It is important for this role to be recognized as an important agricultural activity. There is a tendency to view food cultivation as agriculture and food distribution simply as marketing and so the role of women is hardly recognized and supported. ECLAC (2005) confirms the dominance of women in the domestic marketing and distribution of food. They were the main participants in the huckster trade in the eastern Caribbean traveling among islands in the Caribbean archipelago in the Lesser Antilles trading in agricultural produce. There is an urgent need to put institutional structures in place to better connect women to markets as they are the key players in the food exchange chain. An important element in doing this is creating an atmosphere where food marketing through the produce markets becomes a more dignified economic activity and is recognized as such. This would enhance farm incomes and positively affect household quality of life, education, food security, and nutrition. It would also increase women's self-esteem and empower them to expand their role and efforts. Any successful strategy to reduce gender inequality in agriculture must be based on the principle that there are other agricultural roles outside of cultivation or livestock and poultry rearing and that many women are involved in these other roles.

7.4.3 Women's Groups

There is a need to empower women in agriculture in the Caribbean. Related to this empowerment is the creation of agency among agricultural women in the region. This is critical if women are to take control of the process of their empowerment and play a leadership role in bringing about gender equality in agriculture. An important avenue for achieving this is supporting women in their effort to mobilize and unite toward common goals and addressing common problems faced by women agriculturalists. This can be done through women's agriculturalist groups. Currently there are a number of rural women's groups across the region under the umbrella of CANROP, but women agriculturalist groups should include urban women farmers, because of the vital role they can and do play in household food security and nutrition. Women agriculturalist groups have been successful in promoting collective action and leadership and building capacity of women to take advantage of opportunities to play leadership roles in household food security and income generation.

7.4.4 Research

Generally speaking, there is a growing knowledge base about women's role in food and agriculture and food security and nutrition more specifically. However, in the Caribbean this global trend does not hold true. There is a lack of institutional analysis and academic research in the area of women's role and their real and potential contributions to food security and nutrition and household incomes from agricultural activities. This means that there is a lack of high-quality data and information to guide analysis and policy-making and very little in the way of evidence to support what is known intuitively through educated guess and very-micro-scale research, much of which is quite dated. This lack of evidence about women's contribution feeds into the neglect and indifference toward the efforts of women agriculturalists. Also severely lacking is women's perspectives and active engagement in research on the subject. Research should be geared toward design and implementation of gendered agricultural policies and food security strategies. This should include research on domestic food crops including a focus on vegetable and fruit cultivation for home consumption with micronutrition in mind and appropriate preparation to maximize nutritional benefits. Caribbean populations have become estranged from local traditional foods, and women can play a vital role in research and use of traditional wild edible plants with a wide variety of uses as they do elsewhere (Boothrod, 2010).

There is need for special programs for women in the areas of agricultural extension, agroprocessing, and other value-added activities, agricultural loans to women, education in farming techniques and agrobusiness including agroprocessing and marketing, access to land and secure tenure, access to technology.

7.4.5 Community and Household Agroprocessing

Most of the food produced in the Caribbean is sold as fresh produce. An important component of food security and women's participation should be the initiative to promote increases in agroprocessing at the household or community levels. Agroprocessing would drastically reduce postharvest losses, preserve food, and add value, thus increasing farm incomes. Cottage industries based on locally produced fresh farm produce should be promoted, encouraged, and supported. In this regard there should be efforts to establish properly constituted cooperatives but household-level industries should also be pursued. Women could also play an instrumental role here as they have a rich history with cottage industries in the region. Again there will be a need for training and ongoing learning in areas like food processing, business management, marketing and distribution, and accounting.

7.4.6 Women and Home Gardening

Women's involvement in home gardening has been found to be an effective strategy for household food security and nutrition in many parts of the developing world with most of the literature coming from Asia and Africa. Home gardening is not women's work—research indicates that in Sri Lanka and Indonesia (Hoogerbrugge and Fresco, 1993), Bangladesh (Talukder et al., 2000), and Russia (tho Seeth et al., 1998) women might do most of the work on the subsistence side, but men and other family members are also involved and can have specialized roles (Brownrigg, 1985; Mitchell and Hanstad, 2004). Marsh (1998) suggested that sale from home garden produce may be the only independent source of income for women and could therefore be very important. Other research has shown that operating a significant home garden is a status symbol for women in some places and is representative of women's economic independence, self-sufficiency, and success in managing their families (Finerman and Sackett, 2003). A significant role of women in home gardens can lead to improvements in household nutrition in general and the nutrition of children in particular (Kumar and Nair, 2006; Kumar, 1978; Talukder et al., 2000). This is true in both rural and urban spaces where men tend to focus on commercially lucrative crops while women prefer to concentrate on cultivation for home consumption. Ninez (1985) provides an example of this gendered cultivation orientation in the urban gardens of Lima, Peru.

Because of women's focus on production for home consumption, their control of home garden resources could be an effective strategy for combating malnutrition and micronutrient deficiencies through the production of fruits and vegetables and small livestock and poultry. For example, Hoogerbrugge and Fresco (1993) report that in Sri Lanka, home gardens produce 60 percent of leaf vegetables and 20 percent of all vegetables consumed by households. In Bangladesh a relationship has been found between the presence of home gardens and household health and nutrition. It was found that in households with home gardens there were less incidents of eye diseases caused by vitamin A deficiency than in households that did not have a home garden (HKI/AP, 2003). Households growing more and larger variety of fruits and vegetables in Bangladesh could have higher intakes of vitamin A (HKI/AP, 2001) and fruits and vegetables were found to make a greater contribution to vitamin intake by reproductive-aged women than animal products (Bloem, 1996). In addition, home gardens provide other nutritional benefits. For example, in Java they contributed 18–40 percent of calories and in the Philippines produced almost 60 percent of recommended daily caloric intake. Thus they provide not only vitamins but energy foods as well and are thus more than just fruit and vegetable gardens.

The promotion of home gardens and agrobusinesses related to them among unemployed and underemployed Caribbean women has a lot of potential to impact food security at the household level not only through enhanced food availability but also through the supply of micronutrients. Women should also be engaged through a coherent urban agriculture strategy toward food security. Regional efforts to boost food production, enhance food availability, and improve nutrition will be counterproductive if the agricultural policies continue to ignore the important role that women can play. The marginalization of women farmers places regional food security at risk as much as other factors that are given more attention in the literature such as climate change.

7.5 Summary

From the discussion in this chapter it is clear that there is overwhelming evidence of the critical role women play in agriculture and food production and their contribution to food security and nutrition especially at the household level. We have provided here a very brief overview of some necessary steps toward strengthening women's role in food and agriculture and food security in the Caribbean. Of significance is the role that women can play in enhancing household food security, especially nutrition. This is a hot global trend that is a fundamental element in food security planning in Asia and Africa. Yet the poverty alleviation and food security strategy in the Caribbean includes no special plan for increasing women's role. The problem in Caribbean agriculture is not so much discrimination against women but an underestimation of first of all the important role they play in food production and the untapped potential they hold in enhancing food availability and household nutrition. Failing to invest in women farmers is detrimental to sustainable and more productive agricultural systems and enhanced food security.

CHAPTER 8

Food Forests and Home Gardens: Their Roles and Functions in Domestic Food Production and Food Security in the Caribbean

8.1 Overview

In the Caribbean, kitchen gardens can be categorized into *home gardens* and *food forests*. Home gardens are typically small cultivated areas in the backyard, where vegetables, pepper, sweet potato, pumpkin, callaloo (spinach), a few ears of corn, and other short-term crops are grown; the produce is strictly for home consumption and for sharing with relatives and friends. Food forests are also grown on the home plot. They are low-maintenance and low-input agroforestry systems, characterized by a wide diversity of plant species—predominantly fruit and food trees such as mango; various species of citrus, mainly orange, grapefruit, and tangerine; ackee; breadfruit; avocado; jackfruit; coconut; plantain; and banana—and also nonfood perennial hardwoods such as cedar, mahoe, and mahogany, which are commercially lucrative lumber trees—are grown. Other woody perennials are grown for their ecological benefits to the system. Food forests also have an undergrowth of plants such as medicinal herbs and bushes and food crops such as coco yams, dasheen, sugarcane, and yam. These agrospaces mimic a forest in terms of the vertical configuration of plants of different heights, hence the terminology "food forests." The literature on Caribbean agriculture sometimes refers to these food forests simply as kitchen gardens (see Brierley, 1991, 1976, 1974; Hills, 1988; Thomasson, 1994), but we urge readers to use this description carefully, because to the average Jamaican, for example, a kitchen garden might simply refer to the backyard vegetable plot.

Kitchen gardens are not used just for growing food-producing plants and trees, but also serve important cultural and social functions. According to Richards (1989), for resource-poor farmers, farming involves not only food production but also a continuous experimentation. In this context, the Caribbean kitchen gardening is an ongoing experiment in which farmers use their local traditional knowledge to experiment and innovate as they adapt and manage scarce resources. The kitchen garden approach to household and communal food security is a ubiquitous mainstay of traditional agricultural and subsistence systems in the tropics and is a fundamental element in food security in the Caribbean among urban as well as rural populations. The approach is a clear manifestation of the sophisticated agro-ecological knowledge held by individuals belonging to small-scale farming communities (Brierley, 1991; Hills, 1988; Thomasson, 1994).

The diverse range of species grown in Caribbean kitchen gardens closely mimics those in natural tropical ecosystems, in terms of both structure and benefits; the different rooting systems and canopies are combined to facilitate the sustainable use of moisture, nutrients, and light, just as in natural ecosystems. Caribbean farmers' knowledge of the unique benefits of different crop combinations has been well documented (Brierley, 1991, 1976; Flora, 2001; Gliessman, 1998, 2001; Hills, 1988; Hills and Iton, 1983; Jackson and Jackson, 2002; Thomasson, 1994). The significance of kitchen gardens is twofold: they provide meals to numerous Caribbean families and are also a source of income for them.

In the Caribbean, kitchen gardens are sites of experimentation by the farmers and are considered to be training grounds for children to learn about farming. Rural children, in particular, first learn about and experiment with the mutually beneficial combinations of crops and then learn to distinguish useful medicinal plants and herbs from "bush"; they also learn the nuances of mixed farming, which are ubiquitous to small-scale food farming systems, to be imbibed and refined. In the kitchen gardens, traditional farming practices and local farm knowledge become part of the body of knowledge held individually and in common (Brierley, 1976, 1991; Thomasson, 1994).

Kitchen gardens are sites of sustainable production and resource use that help to preserve ecological balance (Beckford et al., 2007). They are detritus, rich with soil fertility maintained through natural decomposition of vegetation and nitrogen provided by animal waste and leguminous trees and plants (Jackson and Jackson, 2002). Thomasson (1994) describes Caribbean kitchen gardens as an adaptive survival strategy among resource-poor, small-scale farmers. The kitchen gardens are a vital source of food supply for household consumption, but also play an important income generation role for many rural households.

This chapter will discuss the historical and contemporary role of kitchen gardens in individual and household food security and livelihoods in the Caribbean, and will examine strategies for enhancing their contribution.

8.2 Multistory Tropical Food Forest Systems in a Global Context

Kitchen gardens are synonymous with tropical food production systems. Discussions about kitchen gardens can get confusing as they are known by a variety of different names in different parts of the world. Most of the research and literature about them come from the tropics, and as a result, temperate kitchen gardens are not as well known. In Europe—the UK and Russia in particular—they are known as *forest gardens,* while in Sri Lanka, they are called *Kandayan forest gardens.* In south India, Nepal, Zambia, Zimbabwe, and Tanzania, the term *home gardens* is used. In Mexico, similar systems are called *huertos familiars* ("family orchards"). In Java they are known as *pekarangan,* or the gardens of "complete design." Much of Mayan food supply came from their *petkot,* meaning "orchard gardens" (Smith and Masson, 2000). In some African countries, notably Tanzania, Zambia, and Zimbabwe, this type of land use is widespread not only in rural areas but also in urban and periurban locations where they play a critical role in food production and household food security. Among the Chagga people, too, who inhabit the lower slopes of Mount Kilimanjaro, Tanzania these gardens are well known (Smith and Masson, 2000). In parts of North Africa, traditional forest gardens integrate palm trees, fruit trees, and vegetables. In the academic research literature from the Caribbean, these food production systems are referred to as *kitchen gardens* (Brierley, 1991, 1976, 1974; Thomasson, 1994), *food forests,* and *tropical mixed gardens* (Hills and Iton, 1983). Thomasson (1994) also used the term *household food gardens* perhaps to capture their subsistence production purpose and contribution to household food security.

For the purpose of our discussion, we will use the term *food forests* and *kitchen gardens* interchangeably. Regardless of what they are called in various parts of the world, these food production systems are common in the tropics, where mixed cropping is used to cultivate a wide variety of crops and other tree species in what could arguably be described as the most enduring example of intensive farming in the world. According to Nair (1993), these food production systems can be found in almost every tropical and subtropical ecozone that are dominated by subsistence land-use patterns. In some parts of India and in Indonesia, the kitchen garden or home garden, as they are called, is the most common form of land use. Some home garden systems in the Asian subcontinent and sub-Saharan Africa integrate some form of animal husbandry and aquaculture as well.

Because of the various systems that exist throughout the tropics, we will refrain from trying to give a precise definition of kitchen gardens and instead try to give a description of these systems based on their most common characteristics wherever they occur in the tropics. As Nair (2001) and Kumar and Nair (2004) point out, there is no universally accepted definition for such systems. The kitchen gardens we write about in this chapter may be described as multistrata systems characterized by a vertical structure that mimics the structure of natural tropical forests. They are essentially agroforestry systems based on low-maintenance organic food production. Throughout the tropics, wherever a kitchen garden exists, it does so as a multispecies food production system with natural layers of above- and below-ground plants interacting to share the resources available for plant growth. According to Mitchell and Hanstad (2004), the diversity of plant species and a layered canopy are the most distinguishing features of food forests. These forests are generally dominated by crops and other tree species, but sometimes livestock and fish farming are also integrated. Although there is no single definition for such systems, Kumar and Nair (2004) posit that the concept of "home gardens" is commonly understood to refer to integrated multistory combinations of trees and crops sometimes with livestock around the homestead. Using the term *tropical home gardens,* they describe them as one of the oldest forms of managed land-use systems in the world. They observed that these multispecies systems are thought to be exemplars of ecological sustainability as they provide food for millions of people. The sustainability of these systems is manifested in biophysical advantages for example, efficient cycling of nutrients as a result of the vast range of species diversity, conservation of biological and cultural diversity, and potential for gender equality in agricultural systems and households (Kumar and Nair, 2004). Based on their mixed plant structure, these systems would also be expected to bring other benefits that are difficult to quantify, such as carbon sequestration. Despite their long history and continuous contribution to food security, nutrition, and household income, kitchen gardens have not been very well studied, with most accounts consisting of descriptive identification of species diversity and their structure (Nair, 2001; Kumar and Nair, 2004). Nair (2001) argued that the economic value, ecological benefits, and sociocultural importance of kitchen gardens warrant a much closer scientific attention than they have received.

8.3 Caribbean Multistory Food Forests

Multistory tropical food forests in the Caribbean represent traditional land-use systems. These systems are largely influenced by the biophysical and sociocultural contexts in which they exist (Kumar and Nair, 2004). This is true of Caribbean multistory food forests (MFFs), which occupy a critical place in food security and livelihood, the significance of which is perhaps grossly underestimated. The importance of this land-use system in the Caribbean is captured in the following quote:

> Kitchen gardens in the Caribbean are intriguing pieces of real estate, possessing as they do keys to understanding the region's small-farming practices and development. Owing to their ubiquity, limited areal extent and perceived economic insignificance, they receive little more than passing reference in the agrarian literature. They can be considered, nevertheless, to represent a peephole through which insights can be gained of their historical antecedents, the traditional beliefs and practices of farmers, the basis of local diet, and their role in the evolution of small farms. In view of these points, there is perhaps no better starting point from which to gain an appreciation of Caribbean small-farming than through examination of its kitchen gardens.
>
> (Brierley, 1991, p. 15)

Brierley's assessment of the worth of kitchen gardens was based on his research on Grenada in the Windward Islands, located in the eastern Caribbean. This assessment is supported by research in the Leeward Island state of Montserrat:

> [K]itchen gardens in Montserrat show far reaching social linkages and benefits not customarily associated with gardens. Paramount are their roles in supplementing and securing the national food supply, in providing employment substitution and reciprocal exchange opportunities, disaster food, and positive psychological benefits. The gardens greatly enrich the quality of life for their elderly cultivators, are essential to the social fabric of the island, and should be considered for their developmental potential by the local Ministry of Agriculture.
>
> (Thomasson, 1994, p. 20)

Caribbean kitchen gardens demonstrate a highly intensive use of land space with a profusion of crops and other plant species in a pattern seemingly without order. But Brierley (1991) argued that what might at first look like congested confusion does make sense when understood in the context of garden farmers applying their knowledge about crops, symbiotic crop combinations, and ecological requirements. Hills and Iton (1983) support this point, stating that in the Caribbean, the food forest or tropical mixed garden is a demonstration of a sophisticated comprehension by many small farmers of the ecological relationships that form the basis of a self-sustaining food production system. Kitchen gardens have traditionally been considered to be the core of small-scale farming systems in the Caribbean (Brierley, 1991) because of their role in household food security and nutrition, their role in the socialization of children to farming, and their function as "old age pension" for many rural families. Ninez (1986) referred to household food gardens as *lifeboats*, because of their multiple roles and functions.

There are several terms used to describe these land-use systems in the literature on Caribbean agriculture. In addition to "kitchen gardens," backyard garden has been used in research in Jamaica (Barker and Spence, 1988); "dooryard garden" was used by Kimber (1973) in his research in Puerto Rico; Thomasson (1994) in his research in Montserrat used the term "kitchen garden" as well as "household food garden"; while Hills and Iton (1983) referred to them as "tropical mixed gardens" and "food forests." Brierley (1991) posits that kitchen gardens are a time-frozen element of Caribbean agriculture and that their descriptions have remained largely unchanged throughout history, which highlights the durability of traditional land use and cropping practices.

8.4 Origin and Evolution of Food Forests

It is believed that kitchen gardens date back to the slave era and evolved from the small parcels of land—called provision grounds—which slave owners provided their slaves to grow some of their own food (Brierley, 1991, 1976; Thomasson, 1994). Watts (1987) suggested that the indigenous peoples of the region might have engaged themselves in incipient forms of kitchen gardening in the pre-European times. During the slave era, provision grounds could have been adjacent to the slaves' dwellings or elsewhere on the plantations. Drawing on their experiences from their ancestral homelands, slaves intensively cultivated their provision grounds using mixed cropping and intercropping practices. The objective was to make the maximum use of limited land space to secure as much food self-sufficiency as possible. According to Hills (1988), these provision grounds were often "food forests" with the likelihood that they contained permanent food and fruit trees, trees for timber, and other plants arranged in a multi-tiered vertical structure. Hills talks about how these slave provision grounds mimicked tropical rain forests and the ecological benefits of these plots (Hills, 1988).

We argue here that if we were to deconstruct the arguments about the origins of food forests, we might come to a different conclusion. There is no doubt that elements of Caribbean multistrata kitchen gardens originated from slavery. These would include mixed cropping and intercropping based on the knowledge about the symbiotic relationship between plants and crops. However, the multistory vertical structure formed using timber trees, fruit and food trees, and other permanent deciduous is hard to imagine during slavery on slave provision grounds. Bear in mind that the slaves did not own these lands, and slave ownership frequently changed hands, forcing some slaves to move from one plantation to another and indeed from island to another. A traditional and contemporary feature of small-scale farming in the Caribbean is that lands with insecure tenure, for example, leased and rented land, were and are still cultivated with short-term removable crops— permanent tree crops are almost never cultivated on such lands. It is unlikely, therefore, that the slaves would be planting permanent fruits, food trees and trees for timber on their provision grounds—that is, plantation owners' land.

We would suggest, therefore, that it is more likely that multistory kitchen gardens or food forests developed slowly with manumitted slaves (slaves who managed to buy their freedom) and slaves who were the offsprings of unions between female slaves and plantation owners who might have had more secure tenure of land. We believe that the rapid development of food forests occurred after emancipation in 1838, when freed slaves wanting to distance themselves from the plantation and the stigma of slavery left the plantations, established independent settlements, and started the development of the Caribbean peas-antry. These free villages were founded on small-scale agriculture and other informal activities like fishing and charcoal burning. Agriculture became the backbone of island economies and multistory food gardens became the cen-ter piece of the farming system and social status as they were tied to the homestead and home ownership, which was and continues to be a marker of success. The free village movement has been most studied in Jamaica—the largest island with the largest number of slaves who consequently became free persons. Any of the original free villages in the island would literally have thousands of these multistory food forests that have gone through several generations of existence.

8.5   Nature of Caribbean Kitchen Gardens: Species Composition and Diversity

Contemporary kitchen gardens have remained basically unaltered over time (Brierley, 1991) which speaks to the resilience of traditional tropical farming systems. Throughout the Caribbean, they are the same and their description on any given island is representative of the region as a whole (Brierley, 1991; Thomasson, 1994). Floyd (1983) describes Jamaican food forests as

> [s]mall patches of land, . . . cropped more or less permanently throughout the year. A varied assortment of root crops, vegetables, and fruit trees are grown, providing rural inhabitants with a convenient supply of food stuffs com-monly found in the local diet. A few yard fowl are the source of eggs and meat. . . . bananas, plantains, citrus, breadfruit, coconuts, maize, cane, red peas,

beans, tomatoes, pumpkins, okra, yams, cocos, irish potatoes and cassava may all be growing together on the same few square yards of soil.

(Floyd, 1983, cited in Brierley, 1991, p. 19)

Based on our research in Jamaica and our review of the literature on kitchen gardens in other parts of the Caribbean, we have developed a profile of this traditional food production system. The first characteristic is that they are ubiquitous to agriculture in the Caribbean. Second, they are part of the homestead and occupy space at the back, front, and sides of dwellings. Third, they are characterized by high species diversity. Fourth, they make intensive use of space. Fifth, their morphology reflects a multistory canopy and vertical structure. Sixth, they are relatively small, especially the more recent ones. Seventh, they serve both subsistence and commercial functions but most of them provide supplemental source of income and consumption. Eighth, they continue to be a feature of new settlements and homesteads. Ninth, they are also features of periurban and urban/suburban landscapes. Based on our earlier discussion of these systems in a global context in this chapter, we would add to these characteristics sustainable recycling of nutrients and general ecological soundness (Kumar and Nair, 2004).

There are also examples of food forests in Jamaica away from a homestead typically comprised of permanent tree crops—fruit trees and deciduous hardwoods including commercial timber species. These might be similar to what Thomasson (1994) observed in Montserrat. Thomasson writes about two types of household food gardens found throughout the Caribbean—kitchen gardens, which are those immediately surrounding the home, and more distant provision gardens, which he calls mountain gardens and which Pulsipher (1979) found at higher elevations in more mountainous islands. The typical kitchen garden comprises a tilled area where an assortment of vegetables, peas and beans, and roots and tubers are grown and another area of woody perennials dominated by fruit trees and commercial timber species.

Brierley (1991) cites research that shows that there is a strong regional uniformity in kitchen garden plots especially in relation to their morphology—dense horizontal structure and multi-tiered vertical structure (Berleant-Schiller and Pulsipher, 1986; Pulsipher, 1989; Hills and Iton, 1983; Innis, 1961, 1980; Kimber, 1973). Observed differences were attributed to differential environmental conditions and human food preferences. Brierley noted that in Grenada the majority of kitchen gardens had an area of less than 2000 m² with the structure of food forests with trees forming the super canopy, while smaller trees and shrubs and vegetables making up the understory. Most of the kitchen gardens in Grenada were dominated by trees for food and fruit, timber, and shade. Trees were also used to delineate property boundaries.

Kitchen gardens in the Caribbean are characterized by huge biodiversity of plant species. It is not uncommon to find dozens of plant species in a multistory garden. In Grenada, for example, Brierley (1991) documented at least 25 vegetables, fruits, legumes, and roots and tubers, and close to around 30 fruit and food trees alone. Thomasson (1994) also found an impressive diversity of crops and other tree species in Montserrat. His plan of a typical household food garden showed over 40 species of food-producing trees and crops including medicinal herbs. A recent study of kitchen gardens in several Jamaican communities for the purpose of this volume indicated a very high level of species diversity. In a sample of 57 plots in various ecological zones, it was found that gardens had an average of 41 species overwhelmingly dominated by hardwood perennials. Older gardens had much more species diversity—an average of 53, including medicinal plants. Brierley (1991) found that the main commercial crops in Grenada's small-scale farming systems—bananas, cocoa, and nutmeg—were frequently found in kitchen gardens of the island and were grown mainly for household consumption. This is significant according to Brierley, as the knowledge and experience gained from this subsistence system is invaluable when applied to commercial production of these crops.

Our recent study of kitchen gardens in Jamaica confirmed this and also found that while some crops are common in kitchen gardens throughout the island, local environmental conditions and cultural and historical customs of faming systems in different areas influenced the specific tree crops composition of kitchen gardens. Thomasson (1994) found that in Montserrat, kitchen gardens were older and operated by elderly people. In Jamaica we observed many newer kitchen gardens in various stages of development and with varying degrees of species diversity. There were many old food forests that tended to be larger than the newer ones. But there were many new ones being established as more people come to own their own homes. It is a standard practice for kitchen gardens to be developed with mainly fruit and food trees such as banana, plantain, avocado, coconut, ackee, plums, mango, and sugarcane. Many of these are much smaller than the traditional food forests as they are established in urban and suburban subdivisions with very limited yard space. They also tend to have less species diversity in general, although we did find quite impressive biodiversity and intensive use of space in some kitchen gardens in some subdivisions.

8.6 Role of Present-Day Kitchen Gardens

Caribbean kitchen gardens provide a number of benefits, chief among which are household food supply and nutrition, income support, training ground for young children, role in the evolution of farming in the region, disaster roles, exchange, and social networking. Some might add to these ecological benefits through the perceived sustainability of kitchen gardens.

8.6.1 Food and Nutrition

Caribbean kitchen gardens play an important role in offsetting food security at the household level. Because of the tremendous diversity of crops planted and the climate of the region generally, kitchen gardens provide food all year round and there is always something available to harvest. Thomasson (1994) calculated that Montserrat kitchen gardens accounted for 32 percent of the total food-producing land in the country. They therefore play a big role in the economic stability of the rural poor. They are also important to household nutrition regularly providing ingredients for daily meals. As explained by one householder who was interviewed, "At least a couple days a week most of our food comes from the yard. The only times we do not cook from the yard is when we eat only rice. If we are hungry it is because we don't want what is on the trees." In his 45 by 70 feet backyard in upper St. Andrew we found bananas, plantain, dasheen, breadfruit, avocado, mango, ackee, coconut, soursop, sugarcane, pumpkin, pineapple, and jew plum. There was also a small area planted with corn, yams, red peas, scotch bonnet pepper, bird pepper, and okra. He had also grown a variety of medicinal plants and herbs.

These pieces of ground could accurately be called tropical mixed orchards and their nutritional value could be astronomic. Their nutritional value obviously is further enhanced when poultry and small livestock are integrated. Studies on Caribbean kitchen gardens (see Brierley, 1991, 1976; Thomasson, 1994; Berleant-Schiller and Pulsipher, 1986; Hills and Iton, 1983; Innis, 1980) have not explored their role in providing for the nutritional needs of households. Studies from the Indian subcontinent and other parts of Asia, however, have suggested that their home gardens are critical sources of nutrition and important to the health and well-being of women, girls, and young children.

8.6.2 Kitchen Gardens as Sources of Household Income

From the earliest days of their existence, kitchen gardens have played a role in income generation. Much is often made of their subsistence function, and this has perhaps led to the underestimation of their role in household income generation. Many definitions/descriptions of these tropical mixed gardens include the term *subsistence* or make reference to food production for home consumption. The Caribbean literature—like the global context—indicates that for the most part they play a supplemental role in income generation (Brierley, 1991; Thomasson, 1994; Kumar and Nair, 2004). In his study of Montserrat kitchen gardens, Thomasson found that 67 percent of gardeners sold at least some of the produce from their gardens. Kitchen gardens as a source of domestic food production, more generally, play a critical income substitution role for many rural households. The production of food for household consumption reduces the amount of income used for purchasing food, making more funds available for other purposes.

Kitchen gardens also serve nonfood purposes. Most especially, the older ones have timber trees incorporated. This is a significant economic strategy and can bring more income than food crops. Timber trees are seen as money in the bank. In times of crisis they are used as a source of emergency funds. On a plot we visited we saw two cedar trees being cut down. It turned out that a daughter of the family was starting university and the trees had been sold to fund this very important venture. On another plot a tree had been cut and the wood had been made into planks and neatly stacked in storage. Some had been used in home construction and some was being used to make a coffin for a family member who had died, thus saving the expense of a commercial casket, which they said would cost over J$200,000.

8.6.3 Kitchen Gardens as Sites of Training and Experimentation

Caribbean kitchen gardens are seen as the places where children obtain their first lessons about food cultivation and try their hand at various farming activities. Through observations of their parents and grandparents, they learn about crops and cultivation techniques, and starting at an early age, they assist in farm chores (Brierley, 1991, 1976). Through this socialization to farming, children acquire intergenerational agronomic knowledge including knowledge about animal husbandry and poultry rearing (Brierley, 1991). It is perhaps here that children first gain knowledge about the symbiotic relationships among different crops, which have contributed to such success in mixed cropping systems among small-scale food farmers. It is believed that women play a big role in this training of children. Mothers and grandmothers spend a lot of time with young children in their preschool years, and women were generally responsible for tending to kitchen gardens (Berleant-Schiller, 1977; Horowitz, 1967). Children started out watching their female guardians before taking on light duties and becoming involved in tasks such as watering before gradually taking on more responsibilities (Berleant-Schiller, 1977). They would eventually be able to do harder tasks like tilling the soil at which stage they would spend more time on the regular farm with their men folk.

The survival of traditional farming knowledge and practices is related to this process of transmitting farming knowledge from generation to generation.

8.6.4 Cultural and Social Functions

Kitchen gardens play a social and cultural function that is misunderstood and underestimated. Many kitchen gardens in the Caribbean include plants and herbs that have been ascribed medicinal benefits. Home remedies for a wide range of maladies are brewed from home gardens. A cornucopia of plants exist, which are used to make traditional medicines used as cleansers; deworming concoctions; pain killers for toothache, headache, and bellyache; for diabetes ("sugar") and hypertension; and a host of products ("bushes") for pregnant and lactating mothers, stamina, strength, and aphrodisiacs. There are also plants to ward off evil spirits and to prepare baths to wash away bad luck. Unfortunately, because traditional knowledge is orally transmitted and rarely documented, this traditional medicinal knowledge is disappearing as the holders of this knowledge die with no written records of their knowledge.

In Montserrat and Martinique, kitchen gardens are also important in gifting or exchange (Thomasson, 1994). Kitchen garden produce is routinely gifted to visitors to the home as well as to family members and neighbors on a reciprocal basis. It is thus important in community relations and social networking. In Jamaica, rural migrants to urban areas will go to "country" on weekends and long holidays and bring back foodstuff from the old homestead that they often share with their urban friends and colleagues.

Thomasson (1994) also shows how kitchen gardens in Montserrat were important to food security stability in the immediate aftermath of Hurricane Hugo, when food supply chains on the island were severely disrupted. Whereas tree crops were badly damaged by the hurricane, the preponderance of tubers and root crops in the mountain gardens of the island meant that disaster food was available for harvesting in the weeks and months after the hurricane.

8.7 Summary

Caribbean kitchen gardens serve multiple functions. They are sites of initial agricultural training and experimentation. They are efficient food production systems with subsistence and commercial roles. They have significant implications for household food security and are therefore important to food security at the national level as well. A great deal of staple food is produced in kitchen gardens, and we argue that they can play a vital role in production of calories and household nutrition. Kitchen gardens also have critical implications for economic livelihoods, and the proliferation of roadside fruit stands in many parts of the region is a reflection of the importance of kitchen gardens as sources of income. Kitchen gardens have also become a feature of the suburban landscape.

Kitchen gardens are good examples of traditional knowledge at work. They are also great examples of intensive agriculture and sustainable agriculture. Householders have intensified their use of land resources to produce more from the same amount of land. These detritus rich ecosystems are thought to efficiently recycle nutrients to create a balanced ecological system.

We question the extent to which the potential contributions of kitchen gardens are being maximized. Because they have typically been considered to be sites of subsistence production, their contribution to food security, nutrition and health, and general rural development has not been considered seriously in the food security plans. We would suggest that this needs to change. Kitchen gardens are a monumental part of rural food supply, and the marketing of fresh fruits means that their reach extends outside the household. Their productive capacities can be enhanced through research and development and extension services and their income-generating capacity can be improved through the strategic promotion of cottage industries and agroprocessing.

There is still a lot to be learned about Caribbean kitchen gardens. An intriguing question begging to be answered relates to their role in household nutrition and calorie intake. Just as important is their role in income generation.

CHAPTER 9

Urban Agriculture for Food Security in the Caribbean

9.1 Overview

The issue of urban agriculture has become increasingly significant in food security strategy globally as the world's urban population (especially in developing countries) continues to climb rapidly and urban poverty spirals. It is thought that if current trends continue by the year 2020, more poor and undernourished people in developing countries will live in cities than in the countryside (Food and Agricultural Organization, 2001). Following the lead of the FAO, many countries and municipalities are making urban and peri-urban agriculture (UPA) an integral part of their food security and poverty reduction strategy. In 1994 the FAO launched the Special Program for Food Security (SPFS), a multidisciplinary program that integrates the perspectives of many academic disciplines to promote a holistic approach to food security. This approach was endorsed by world leaders at the World Food Summit in 1996 (FAO, 2001). A major component of the SPFS design is an element of UPA aimed at improving access to food for people living in (urban) and around cities (peri-urban).

Urban and peri-urban agriculture may be described as the production, marketing, and distribution of agricultural and food products within city cores (urban) and on the outskirts of cities and towns (peri-urban). Activities occur in different locations and at different scales and might comprise home gardens/backyard and frontyard gardens, school gardens, community gardens, roof tops, balconies, open spaces, along river and gully banks, and along railway tracks (FAO, 2001). Activities might include the growing of crops, livestock and poultry rearing, aqua-culture, and charcoal burning and typically serve both subsistence and commercial purposes. UPA is basically food production by and for the poor (FAO, 2001). The FAO says that in most cases, urban agricultural production is an adaptive response of the urban poor to inadequate, unreliable, and irregular access to food (FAO, 2001). This might be due to lack of availability of food or entitlement issues, that is, inability to purchase food.

The benefits of UPA include improved food and nutritional security for the urban population, especially the urban poor; employment; and income generation (Pasquini and Young, 2009). Environmental benefits have also been identified. For example, it is argued that if properly managed, large quantities of urban waste could be composted for UPA, recycling soil nutrients and enhancing soil quality, and alleviating waste disposal challenges that face so many cities in developing countries (Cofie et al., 2003, cited by Pasquini and Young, 2009). UPA can also play a pivotal role in mediating individual and household livelihood vulnerability (Ambrose-Oji, 2009).

There is a growing body of research on the dynamics and potential of UPA in some parts of the tropics, especially parts of Africa and Asia, but there is a dearth of scientific research on this phenomenon in the Caribbean. UPA does occur in the Caribbean, but its significance and potential is not widely known and its various forms have not been adequately researched.

In this chapter we will examine the role and potential of urban agriculture in regional food security. We will look at examples of urban agriculture and discuss ways of promoting UPA in the region. We offer the perspective that UPA should become a significant part of the overall food security strategy in the Caribbean. It is important to note that while it is possible to distinguish urban from peri-urban agriculture, in this chapter we treat them as similar in roles, functions, and potential.

9.2 Urban Agriculture in the Global Context

Large cities have existed throughout history, but the growth of cities over the past 60 years has been spectacular. The real phenomenon has been the growth of cities with over 10 million people—aptly referred to as mega-cities. In 1970 there were five such cities in the world, with three in the developing world. It is projected that by 2015 there will be some 564 cities with at least 1 million people—425 of these will be in developing countries (Mougeot, 2006). The urban population is therefore growing at a rapid rate, but the number of urban poor is rising even more quickly (FAO, 2001). Urban poverty is a predictable result of the burgeoning urban population. There is a high rate of unemployment and underemployment, and a large and growing segment of the urban population resides in unhygienic conditions such as slums and squatter settlements without water and sanitation. UN-HABITAT (2004) estimated the rate of global urban poverty to be 30 percent and that it would grow to 50 percent by 2020. Nearly all the increase is expected to be in the developing world. A major manifestation of urban poverty is hunger and food insecurity. The FAO (2001) estimated that by 2020 more poor and undernourished people in developing countries will live in cities than in the countryside. Mougeot (2006) posited that food has become a "basic luxury" for the urban poor. Urban households can spend as much as 80 percent of their income on food (PCC, 1990). Malnutrition is rampant in the poorest parts of the developing world, where one meal per day is not just common but typical.

Many urban dwellers are therefore making an effort to supply some of their food needs themselves. According to the FAO (2001) up to two-thirds of urban and peri-urban households in developing countries are involved in agriculture. The UNDP estimates that 15 percent of the world's food is produced by some 800 million urban farmers (UNDP, 1996). In general, urban food production is a response to food insecurity caused by inadequate access to food on a sustainable basis due to the lack of availability or issues with entitlements.

9.2.1 Definition and Description of Urban Agriculture

"Urban agriculture is an activity that produces, processes, and markets food and other products, on land and water in urban and peri-urban areas, applying intensive production methods, and (re) using natural resources and urban wastes, to yield a diversity of crops and livestock" (UNDP, 1996). The FAO-COAG (1999) adds that UPA is located around and within cities competing for resources, which could serve other purposes to satisfy the needs of the population. Mougeot (1999) describes UPA as the growing, raising, processing, and distribution of a diversity of food and nonfood products reusing materials and resources found in these areas and supplying products to these areas. It is believed that UPA is a growing phenomenon in developing countries in response to increasing urban poverty (FAO,2001; UNDP, 1996; Mougeot, 1999). Urban agriculture goes beyond food production, and the broad spectrum of activities used to achieve household food security and income must be considered if it is to be fully understood (Jacobi, Drescher, and Amend, 2000). In its various forms whether it is community-based or individual household food production, UPA serves many purposes to city dwellers.

Urban food production activities meet many needs. The following list by the FAO provides a useful summary of these:

1. Aid food production, income generation, and recreational opportunities
2. Contribute to the prevention of micronutrient deficiency

3. Can enrich the urban environment
4. Improve access to consumer markets
5. Imply less need for packaging, storage, and transportation of food
6. Create potential agricultural jobs and incomes
7. Provide nonmarket access to food for poor consumers
8. Increase availability of fresh, perishable food
9. Improve proximity to services including waste treatment facilities
10. Create opportunities for waste recycling and reuse possibilities
11. Contribute to preserve and improve biological diversity by integrating it in the ecosystem

(FAO, 2001, p. 12)

Nugent (1997) identifies other benefits of urban agriculture, including provision of local jobs, making cities green and beautiful, empowering urban people, improving freshness and variety of produce, and involving people in the production of their own food.

There has been considerable progress in cities of the developing world mainly in Africa in the past decade and a half or so (for example, in Dar es Salaam, Tanzania, nearly 70 percent of families are involved in agriculture) but weaving agricultural land use into the policy and practice of the urbanscape has been challenging. Where urban agriculture exists in parts of the developing world, it is still largely unregulated, haphazard, and coincidental. The Caribbean, with the exception of Cuba, is an example of this.

In terms of land area, urban agriculture is practiced on small- to medium-size plots in the city. Peri-urban agriculture occurs on larger spaces on the outskirts of cities. Due to the open space constraints, which characterize many large cities, most urban gardens tend to be small whether they are household gardens or community gardens (though the latter tend to be larger). A common element of scale is the urban micro-gardens, which are very small and intensely cultivated areas typically found in the heart of the urban core. They are simple, low-cost, and low-technology systems and are tailor made for the high-density urban core and suburbs (FAO, 2001). Their size and function facilitate their establishment in very small spaces including backyards, roof tops, balconies, and road sides. Hydroponics is becoming popular in many urban agricultural systems as well. They are successfully utilized by the urban landless poor to produce a diverse range of mainly vegetables, herbs, and spices for home consumption (FAO, 2001; Jacobi, Drescher, and Amend, 2000). Micro-gardens are thought to be ecologically sustainable cultivation systems using mainly recycled materials and creating a closed system with high efficiency in water use (FAO, 2001). Urban micro-gardens tend to be completely organic and use of synthetic fertilizers and chemical pesticides is the exception rather than the rule.

9.2.2 Characteristics of UPA Farmers

The majority of urban farmers are low-income earners who grow food primarily for household consumption but also engage in commercial practices (Jacobi, Drescher, and Amend, 2000). This is a very informal activity typically without institutional support or recognition. Both men and women are involved, but some activities such as home gardening, poultry rearing, and small livestock rearing tend to be dominated by women. Subsistence production is typically the role of women with more men engaged in commercial urban food production (Jacobi, Drescher, and Amend, 2000; Mougeot, 1999). Mougeot (1999) reported cooperation between urban farmers engaged in different production systems as they used the same tracts of land at different times for different purposes and engaged in product and resources exchange. It should be noted that urban agriculture cuts across income groups even though it is dominated by low- and medium-income people.

9.2.3 Role of Women in Urban Agriculture

Women and children are the most marginalized groups in the urban population, and not surprisingly, play a big role in urban agriculture (FAO, 2001). Single, elderly, and widowed women are especially marginalized. The FAO argues that women tend to dominate urban agriculture because they are disadvantaged in other forms of employment in the formal sector (2001). In some areas of Africa, urban agriculture and women's farming are synonymous, indicating the predominance of women's participation. In many cities, urban agriculture has become a niche area of economic activity and employment for women (FAO, 2001). Women play a critical role in household food supply. It is believed that income earned by women has a greater positive impact on a family's health and nutrition than men's income (Kumar and Nair, 2001; Mougeot, 2006).

Urban agriculture is attractive to women as it provides flexibility and allows them to work close to home (FAO, 2001). They are able to provide extra food, which is typically safer and contributes more to household nutrition than store-bought food (FAO, 2001). There are, however, limits to women's participation in urban agriculture. Women face challenges in terms of access to resources and services (Mougeot, 2006; FAO, 2001). In some areas, they have less access to education as well, and cultural barriers, laws, and customs may place limits on their ability to own assets and participate in decision-making. Additionally, the commercialization of urban agriculture is adversely affecting the ability of women to participate in agriculture (FAO, 2001). But as Mougeot (2006) concludes, women find ways to beat these odds and engage in successful urban agriculture and often dominate the distribution side of urban food production where they function as middle traders buying from producers and reselling or processing fresh produce.

Urban agriculture is practiced in both developed and developing countries. In developed countries, most urban farmers are upper or middle class folks who engage in urban agriculture activities as a hobby or as an enjoyable and rewarding pastime. In some communities in developed world cities, urban agriculture is promoted through community gardens as a way of empowering marginalized people and building self-esteem. In the developing world, the motivation for urban agriculture is far more functionally related to fulfillment of the most basic of human needs—that is, food.

9.3 Urban Agriculture in the Caribbean

Globally the practice of urban agriculture has increased rapidly over the past 30 years. It is believed that over 800 million people all over the world are growing crops and rearing livestock to supplement their food supply and earn income (FAO, 2001). In many parts of Africa, Asia, and Latin America, governments are weaving urban agriculture into the land-use policy of cities or at least encouraging urban dwellers to become involved or more involved in food production activities. This has not always been the case, and indeed in some places, agriculture is still seen as a rural or "countryside" activity not suitable for urban environments (FAO, 2001). Especially as it relates to livestock husbandry, there is the fear that competition for land and water and agriculture competing with more valuable land uses could cause problems. In some places, urban agriculture was popularly associated with poverty and underdevelopment. It is unclear as to whether this mindset plays a major role in Caribbean thinking, but it is clear that urban agriculture has just recently began to be taken seriously in the region and still does not have meaningful legislative or policy support. The one notable exception is Cuba, which, as we discuss later in this chapter, is a world leader in the area of institutional support for urban agriculture.

The status of urban agriculture in the Caribbean is epitomized by the lack of research and literature on the topic in the region. Again with the exception of Cuba, case studies of urban agricultural praxis in the Caribbean are lacking. This does not mean that urban agriculture does not take place. In fact, it is common in both urban and peri-urban spaces in many big Caribbean cities—Kingston, Montego Bay, Kingstown, Bridgetown, Georgetown, and Port of Spain. It is also common in smaller towns across the region. Where it exists, however, it still tends to be a fringe activity, small-scale and informal, unrecognized and unsupported. Urban agriculture in the region has not been a major part of the discussions about poverty and development and food security initiatives have overwhelmingly focused on rural spaces which present a certain irony, given the marginalization of the rural, generally speaking.

9.4 The Role That Urban Agriculture Can Play in the Caribbean

9.4.1 Food Security

Urban agriculture contributes to the availability of and access to food year round for the urban poor (FAO, 2001; Mougeot, 2006). It also contributes to safer food consumption given the organic nature of the food produced. Urban agriculture therefore contributes to urban food security, and the production of many staple foods is a safety valve in times of food crises. For example, in Soweto, South Africa, urban agriculture was found to be an immediate effective response to hunger and malnutrition (FAO, 2001). It also enhances stability of urban food supply due to year-round production in most cases. The poorest households are benefited as they produce some of their own food, thus saving precious income for other household purposes (FAO, 2001). Through its contribution to the quantity and quality of food intake, urban agriculture therefore contributes to urban food security.

9.4.2 Food Diversity and Nutrition

Urban agriculture is characterized by a great diversity in activities as well as products (FAO, 2001; Mougeot, 2006; Jacobi, Drescher, and Amend, 2000). Activities could include horticulture, vegetable and fruit cultivation, small livestock husbandry, poultry rearing, and aquaculture. In addition there is a growing engagement of small-scale food processing and cottage industry–type value-adding activities. Fresh produce includes staple foods, vegetables, fruits, herbs and medicinal plants, meat, eggs, and fish. The quality of urban diets is enhanced by this diversified range of activities and produce, which contributes significantly to a balanced diet fortified by energy foods, vitamins, proteins, fats, and other essential micronutrients (FAO, 2001). These foods can be prepared in a number of ways, adding to the social experience of food and adding more variety to the mix.

9.4.3 Urban Agriculture and Poverty Alleviation

Urban agriculture can contribute to urban poverty alleviation, income generation, and employment, and can reduce the amount of income spent on food by producing some for their own needs (FAO, 2001; Mougeot, 2006). This is receiving increasing attention by policy-makers in some regions as urban agriculture appears to expand during economic crises (Mougeot, 2006). Income is generated through the production, processing, and marketing of produce. Through the production of food, urban gardeners save money, which is then spent on nonfood items and foods they cannot produce themselves. With an estimated 800 million people involved globally, urban agriculture also generates employment leading to increased income and poverty reduction (FAO, 2001).

9.4.4 Environmental Benefits

Urban agriculture can provide significant environmental benefits as well. It is generally based on sustainable practices especially in relation to the limited use of agrochemicals (FAO, 2001). That said, it should be noted that peri-urban agriculture can have a good deal of pesticide and chemical use as it tends to be more market oriented. In some cities, urban agriculture is also used to create green zones integrating trees, wetlands, and grassy areas. Urban agriculture, therefore, greens and beautifies the otherwise sterile concrete environment of cities. Such areas are believed to have significant benefits for the quality of life of urban dwellers. The recycling and reuse of resources and materials also bring environmental benefits (FAO, 2001; Mouget, 2006).

There can be serious environmental and health risks associated with urban agriculture. For example, plant and animal waste can cause problems, livestock and poultry farming can lead to diseases and bad smells, and soil and water can become contaminated. When food is grown on former dump sites, the presence of contaminants and toxic materials poses a serious threat to health and well-being.

9.5 Characteristics of Caribbean Urban Gardens

Urban gardens in the Caribbean are typically small, highly intensive cultivation systems that use localized irrigation methods. Most watering is done by hand, with a makeshift watering can or a garden hose connected to a potable water supply. Vegetable seeds are sown in small raised nursery beds in the corners of properties. Old containers and old motor vehicle tires are also used to sow seeds before they are transplanted in prepared garden beds, which are usually raised. Large tractor tires are sometimes used to serve as raised beds to grow vegetables. This was noticed in several backyards in Spanish Town, Jamaica, where householders explained that the soil in their yards was of poor quality and so they set up tires and other vegetable home-made planters and fill them with better soil from elsewhere. These are then used to grow vegetables, herbs and spices, and legumes. The most commonly reoccurring crops grown in core urban gardens in the Caribbean include vegetables (tomato, callaloo, cucumber, pumpkin, lettuce, carrots), peas and beans, fruits (melon, banana, plantain), herbs and spices (mint, thyme, skellion, sweet peppers, and hot peppers), and staple foods mainly yams, sweet potatoes, cassava, and corn. In the suburbs, gardens tend to have a much greater diversity of plants in multistory tropical food garden settings with a combination of food crops, food, and fruit trees. For example, in Jamaica, suburban gardens would have fruit and food trees such as ackee, avocado, mango, papaya, soursop, breadfruit, banana, plantain, coconut, and Jamaican apple in addition to vegetables, condiments and legumes, and staples like potatoes, yam, cocos (also called coco yams in West African countries), and dasheen.

It was observed that in the urban core, the kinds of food planted were associated with the purpose of the farm. Some crops like lettuce, carrots, and cucumber (eaten in salads) were grown for sale as they are not a major part of the daily diet of the urban poor except carrots, which might be used in soups. Food for subsistence was grown at home while food for sale was typically grown outside the yard in adjacent open areas or further afield along gully sides and river banks. In Georgetown, Guyana, some urban gardeners have been allowed to grow food on open lots belonging to private residents to prevent the vacant lots being used for dumping. In Kingston, Montego Bay, and Spanish Town in Jamaica, there have also been reports of land owners allowing vacant lots to be farmed to prevent squatting.

Micro-gardens are a common feature of urban agriculture in the Caribbean. The most popular are the backyard gardens, balconies, and roof gardens. Backyard gardens are used to grow a wide assortment of vegetables, legumes, and staple crops. On flat roof tops in Kingston, Jamaica, and Kingstown, St. Vincent and the Grenadines, pineapples are cultivated like potted plants. In some cities of the Caribbean, community gardens are being established. These are used mainly for growing vegetables and flowers. In some communities in Kingston, Jamaica, urban agriculture is popular among members of the Rastafarian faith perhaps due to their desire for more natural and organic foods.

Livestock rearing is not common in the urban core as small spaces do not allow for animal husbandry. Land-use policies do not encourage this, and some prohibit it through strict zoning regulations. There are cases of chicken coops being set up, which can lead to conflict as they are often smelly. Goats are often seen roaming around cities. In peri-urban areas, livestock rearing is much more common and include larger animals like cows, pigs, and goats. Poultry rearing is also common in peri-urban spaces.

Urban agriculture in the Caribbean is associated with informal and localized distribution of fresh produce, which is manifested in the mobile and temporary fruit stands illegally set up on street corners, bus stations, and busy intersections in major towns and cities.

9.6 A Caribbean Example: Urban Agriculture in Cuba

In contrast to other governments in the Caribbean and most across the world, the Cuban government has actively supported urban food production, and Havana, the capital city, is the heart of the urban agriculture revolution. According to Pinderhughes (2000), Cuba is the only country in the world with an extensive state-supported infrastructure to develop urban agriculture. Murphy (2000) says that Cuba has one of the most successful urban agricultural programs in the world. This was precipitated by the food crisis in the early 1990s but manifested in the Cuban philosophy that food was a basic human right (Pinderhughes et al., 2000). According to Murphy (2000) urban agriculture was almost nonexistent in Havana prior to 1989, as even the poorest people had no need to grow their own food because of the effectiveness of the national food distribution mechanism. In the aftermath of the Cuban Revolution, a system of national food distribution was implemented to address hunger and poverty. Havana actually had laws that prohibited the cultivation of agricultural crops in the frontyards of city homes (Murphy, 2000). The Cuban Government's food distribution system was dependent on imported food that came almost exclusively from the Soviet Bloc. The fall of communism in the late 1980s effectively made this source of food unavailable to Cuba, and food shortages became a major concern in the early 1990s. The situation was exacerbated by the United States trade embargo and specifically the Toricelli Bill passed by the Congress in 1992, which made it illegal for foreign subsidiaries of the United States to trade with Cuba. Murray (1993) wrote that 70 percent of this trade was in food and medicine. By the start of 1993, nation-wide food shortages were being experienced with major cities such as Havana being the hardest hit (Murphy, 2000).

The deepening food crisis led to the spontaneous response of the city's inhabitants to begin food cultivation in backyards, on rooftops, in balconies, and in vacant lots adjacent to their homes, growing vegetables and raising small livestock and poultry (Pinderhughes et al., 2000). The great enthusiasm about gardening was not matched by gardeners' knowledge and expertise as the majority of gardeners in Havana were unfamiliar with the agroecological techniques that successful urban gardening required (Murphy, 2000; Mougeot, 2006; Pinderhuges et al., 2000).

Through the Ministry of Agriculture, the Cuban Government responded by establishing a comprehensive, urban food production strategy incorporating land reform, technical assistance, and research and development. Its first step was the creation of the Agricultural Department of Havana in 1994 (Cabada, 1995), which then created the Urban Agriculture Department mandated to put the city's open land into food production and provide resources and services such as extension services, education, seeds, and tools (Murphy, 2000). User rights were secured for urban gardeners and vacant lots—including privately owned lands—were formally designated as spaces for food production (Murphy, 2000; Pinderhughes et al., 2000). Where private land was turned over to would-be growers, they were obliged to cultivate it. If it remained uncultivated for six months, it was returned to its rightful owner (Murphy, 2000; Pinderhughes et al., 2000). Thousands of gardens were established and hundreds of vacant lots were transformed into food production sites largely due to the impetus provided by the political directorate (Pinderhughes et al., 2000).

Most of the gardens were established in areas with high percentage of open land in Havana's outlying neighborhoods where larger spaces for gardens were available. In the older, most densely populated parts of the city, urban agriculture had not been planned for, which resulted in little land for farming (Murphy, 2000). Micro-gardens on roof tops and balconies were attempted but the old roofs were in many cases unsafe. Residents in this area who wanted to garden were given plots in other areas, which required them to travel a bit farther from home (Murphy, 2000). By 1998 there were over 8,000 officially recorded food production units cultivated by over 30,000 people (Murphy, 2000).

In 2000 about 30 percent of the available land in Havana was under cultivation and Havana's urban gardens produced a tremendous variety of food (Murphy, 2000). In terms of crops, 19 main species of vegetables, 21 main species of fruits, 5 main species of roots and tubers, and 5 main species of legumes are cultivated. Rice and sugarcane are also grown (Murphy, 2000). This diversity of food provides all the major food groups and nutritional requirements for healthy eating year round. Urban agriculture has not met Havana's total food needs, but it has raised the quality and variety of food in Havana, and there is a direct link between eating a variety of foods and being healthy (FAO, 1995). Crops are grown organically for the most part with the passage of a law prohibiting the use of chemical pesticides on crops within city limits and providing education on ecologically sound alternatives. Fuster (1997a, cited in Murphy, 2000) reports that as a result of the ban on certain chemicals, Havana's urban vegetable gardens have become the most organic sector of Cuban agriculture. A key aspect of this integrated approach is the nationally coordinated biological control program that promotes prevention of chemical pesticides and use of locally produced pest management agents and has been instrumental in the Cuban survival of the food crisis (Moore, 1997).

A major piece of the urban agricultural strategy in Havana is the network of agricultural extension workers established by the Urban Agriculture Department. Murphy (2000) explains the roles of extension personnel as organizers, teachers, and referral experts who direct gardeners to the appropriate places to acquire information and assistance. They also facilitate interaction among gardeners and inform them about educational opportunities. While there is still work to be done, extension workers have also been very successful in convincing Havana's urban gardeners that large amounts of food can be produced with low external inputs and that pesticides should not be used on fresh vegetables in urban settings (Murphy, 2000).

As part of the urban agriculture strategy, 12 *seed houses* have been established in various parts of the city to sell garden inputs such as tools, seeds, locally produced control agents, biofertilizers, packaged compost, and worm humus, among others (Murphy, 2000). Extension agents work closely with seed houses in order to support urban farmers. Cuba also has an extensive agriculture research sector that facilitates integration of research and development into the agricultural strategy (Murphy, 2000). The strategy also focused on marketing, with the establishment of farmers markets. These have served to eliminate middlemen and allow farmers to sell directly to consumers, resulting in food being more affordable (Murphy, 2000) but perhaps safer too as it cuts down on the number of times the food changes hands.

Urban agriculture has improved the Cuban diet. Havana's gardens were instrumental in relieving the food crisis in the early to middle 1990s and have contributed to quantitative and qualitative food availability in the city (Sanchez, 1997, cited by Murphy, 2000). Food production in the city has relieved the pressure on rural areas to feed the country. Of course, if we were to apply a livelihood perspective, the question that begs to be asked pertains to the impact that successful and expanding urban agriculture has had, is having, and might have on rural agriculture and farm families and rural development more generally. To date the literature is silent on this issue and this is an area in need of research. Urban gardens stimulated local food self-sufficiency, increasing community food security and making food more available and affordable. Gardens are also important sources of food for institutions, which rely on food donations. Schoolchildren especially benefit from this, and it is estimated that 80 percent of urban gardeners donate food (Moscow, 1995, cited by Murphy, 2000), most of which goes to schools and day-care centers.

Cuba's food crisis in the early 1990s became known as the Special Period. Before this the typical Cuban diet did not include a lot of fruits and vegetable. Today things are different and the foods grown by urban gardeners have become important complements to the traditional diet of rice, beans, and meat (Murphy, 2006). The gardens also provide abundant quantities of traditional medicinal plant products.

It was feared that the urban agriculture revolution would lose steam when the food crisis was averted. This has not proven to be the case, however, and Havana's urban gardens have shown steady progress in increasing numbers, size, and quality. They are also helping to bring back traditional foods, which were absent from the diet and local cuisine for years. Key elements of their success have been political will and infrastructural support, access to land, coordination of local resources, support for small producers, strong extension services, demand for fresh produce, and affordable inputs (Murphy, 2002).

It should be noted that urban agriculture has not developed free of problems. There are challenges in maintaining and expanding urban food production in Cuba. In Havana the main problems include water shortage, poor soils, pest and disease control, lack of seed diversity, labor shortage, aging of the urban farm population, and lack of youth involvement (Murphy, 2002). There is a strong sustainability emphasis in the search for solutions to these problems, and building human capacity through education, training, and focusing on prevention is a key component of the strategy (Murphy, 2000).

9.7 Summary

Urban agriculture has tremendous potential to transform cityscapes in the Caribbean and impact the quality of life in a positive way. It can provide benefits related to enhanced urban food security, urban poverty alleviation, protection and restoration of urban biodiversity, and improved urban environment and health management. Urban agriculture can play a major role in the area of gender equity, which is critical to social and economic development. The increased attention by governments to urban agriculture in other parts of the developing world suggests that it is an idea whose time is here. Urban agriculture has been included in food security strategies fashioned by the FAO, UNDP, and other global organizations that underscore its potential. For example, the FAO's Special Program for Food Security underlines the importance of UPA. The FAO Director General requested that Phase One of the SPFS include at least five sites for UPA around big cities out of the 30 sites in this phase (FAO, 2001).

The CARICOM can learn a lot from the Cuban urban garden experience. The totalitarian system of government no doubt allowed expropriation of unused urban land for agricultural purposes, which would not be possible in the liberal democracies of the region, but the promotion and support of urban agriculture would not require appropriating private lands in the Caribbean.

What is needed is government leadership in promoting urban gardening as a dignified activity and an effective way to provide affordable, high-quality, and safe food to the urban poor. Caribbean governments should develop a coherent regional urban agriculture strategy for the cities of the region. There are several essential components that should be included in this, drawing on the Cuban example.

1. They should recognize the significance of the contribution of urban agriculture to social development approaches, generation of jobs and income, self-esteem, enhanced food security, and enhancement of urban environment.
2. They should promote urban agriculture in their cities.
3. They should make land available for urban agriculture with appropriate user rights.
4. They should develop tax incentives for city dwellers involved in urban agriculture activities.
5. They should promote and fund research and development in urban agriculture.
6. They should collect and incorporate data about urban agriculture activities in their planning processes.
7. Urban agriculture should be provided for in the land-use policies of new urban areas.
8. Infrastructure and services should be developed for urban gardeners.

In addition to promoting household and individual urban gardens, governments should also promote and support community gardens and school gardens. Special efforts should be made to increase the participation of women in urban agriculture to improve the nutritional status of their children and families, and empower them through employment and income-earning opportunities leading to economic independence. Before any of this can happen, however, the potential of urban agriculture must be recognized. There is good reason to doubt whether this has happened in the CARICOM region. The recent Regional Food and Nutrition Security Policy formulated by the countries of the region does not make a single reference to urban agriculture as a central strategy.

PART III

Environmental Change, Constraints, Vulnerability, and Survival in Small-Scale Food Farming Systems in the Caribbean

PART III

CHAPTER 10

Climate Change, Hazard Vulnerability, Food Production, and Food Security in the Caribbean

10.1 Overview

Food security in the Caribbean region cannot be discussed without considerable attention to hazard vulnerability. The Caribbean region is prone to meteorological hazards and other natural disasters such as earthquakes and volcanoes. While the region is prone to extreme seismic events, meteorological hazards occur most frequently and cumulatively cause the most hardship. The region is prone to a host of such hazards with hurricanes getting most of the attention, but droughts, floods, and landslides are all features of Caribbean life. In the past 10 years there has been billions of dollars of agricultural losses due to hurricanes and tropical storms. In Jamaica, for example, a cycle of storms caused losses of J$993 million in 2005 alone (McGregor et al., 2009).

Hazard impact is difficult to quantify as there are a number of extreme natural events that might be at work in a given year. For example, between August and September 2005, in Jamaica, severe droughts were experienced the first four months followed by Hurricanes Dennis and Emily and tropical storm Wilma (McGregor et al., 2009). This was just one year in a period characterized by a series of extreme events between 2002 and 2007, with debilitating impact on the small-scale farming sector and agriculture in general. The most extreme of these occurred between 2004 and 2007, beginning with Hurricane Charley and Hurricane Ivan in 2004 (McGregor et al., 2009). This was followed by a seven-month drought, which began in late 2004 and extended well into 2005 (McGregor et al., 2009). This extended dry spell spawned several devastating bush fires that ravaged farms. The 2005 season brought more devastation with the passage of tropical storm Wilma and Hurricanes Dennis and Emily. While 2003 and 2006 were relatively quiet, the vulnerability to meteorological hazards was again emphasized when Hurricane Dean struck the island followed by flood events triggered by two weeks of rain caused by the passage of tropical storm Noel (McGregor et al., 2009). These events severely disrupted the lives and livelihoods of thousands of small-scale food farmers across the country. Other islands across the region suffered as well during this period. For example, the banana industry in the eastern Caribbean was badly affected. In Saint Vincent & the Grenadines, farmers lost their entire banana crop and the spice industry in Grenada suffered significant damage.

There is evidence to suggest that climatic variability is on the rise in the Caribbean and that global climate change is affecting climate and weather patterns in the region. These changes are manifested in the frequency and intensity of weather phenomena and the general unpredictability of climate and weather in the Caribbean. This chapter discusses the relationship between climate change, extreme climate, and weather events and food production and implications for food security in the Caribbean.

10.2 Climate Change

Climate change is likely to impact agriculture and food security across the globe (Slingo et al., 2005). Many agroclimatologists are pessimistic about the effects of climate change on world agriculture. However, some scientists remain confident that humans can and will adapt agricultural practices so that climate change impacts will be minimal or even beneficial. Parry et al. (2004) postulate that the biophysical effects of climate change on agricultural production will be positive in some agricultural systems and regions, and negative in others, and these effects will vary through time. Aggarwal and Mall (2002) remain optimistic and argued that increases in productivity can result from some of the effects of climate change.

10.2.1 Implications for the Caribbean

The susceptibility of most of the Caribbean region to natural hazards such as hurricane, drought, and extreme rainfall events implies that climate change is a cause for serious concern. Of particular significance is the augmentative link between climate change and the frequency and magnitude of hurricanes (IPCC, 2007). For example, the Caribbean Disaster Emergency Response Agency (CDERA) reported that between 1978 and 1988, the Caribbean region experienced only three hurricanes of category 4 magnitude, while the period 1989 to 2001 witnessed a total of 11 (CDERA, 2002). In terms of the magnitude and frequency of hurricanes, Shapiro (1982) notes that an increase in sea surface temperature of 1.5 degrees celsius in the region would lead to an increase in annual average hurricane frequency of about 40 percent (from about 4 per year to between 5 and 6 per year, based on data since the early 1990s).

Brown (2005) observes four possible effects of climate change in the Caribbean. These include changes in rainfall patterns, increased sea surface temperature, increase in natural disasters, and more intense coastal erosion. Because the region is primarily dependent on tourism and agriculture, these possible effects have far-reaching consequences for sustainable development. Slingo (2005) predicts that one associated impact will be a general decline in the productivity of certain crops grown within the tropics even under best-case climate change scenarios.

An analysis of the long-term climatic data relating to these extreme climate events has indicated two major trends within the region. First, the climate of the Caribbean region is changing (IPCC, 2001; Peterson et al., 2002). Specifically, the region has experienced a rise in surface air temperature that is in excess of the global average. Peterson et al. (2002) found that the maximum number of consecutive dry days was decreasing and the number of heavy rainfall events was increasing. Second, statistical analyses indicate that there has been an overall steady increase in the number of hurricanes each year and that this increase may be expected to continue so that the average number of hurricanes per year will increase from 8 to between 11 and 12 over the next 100 years (Peterson et al., 2002). These changes could seriously increase the extent of rural poverty in the Caribbean, increasing vulnerability and inflating food insecurity.

Climate change projections for the Caribbean are causes for serious concern. The region has experienced a rise in surface air temperature that is in excess of the global average. Peterson et al. (2002, p. 8) note that

the extreme inter-annual temperature range is decreasing; the number of very warm days and nights is increasing while the number of very cool days and nights is decreasing.

Gamble (2009) points out that the Fourth Assessment Report of IPCC (2007) supports the trend toward higher regional temperatures and, in reviewing the literature, notes that the rate of increase is not consistent across the region, with some locations showing short-term decreases in temperature.

The aforementioned positive correlation between sea surface temperature and hurricane intensity is not convincing enough to some scientists, who strongly believe that natural cycles of ocean circulation are responsible for the amount and intensity of tropical cyclones in the Atlantic basin (Kossin et al., 2007; Swanson, 2007). For example, there was no statistically significant correlation found between sea surface temperatures (SSTs) and average tropical cyclone intensity in either ocean basin during the 1950–2005 period (Swanson, 2007). The aim here is not to try to settle these debates, but rather to submit an understanding of how local farmers are experiencing these changes and the implications they have for domestic food production and food security.

Hurricanes are the most prevalent meteorological hazards that occur in the Caribbean (Poncelet, 1997; Pielke et al., 2003). The global hurricane belt includes all tropical oceans between latitudes 40 degrees south and 40 degrees north except the southern Atlantic. However, some researchers suggest that

> all portions of Latin America (including Central America and South America) south of 10°N had a less than 1 percent chance of a hurricane strike per year. The annual likelihood of hurricane activity increased farther from the equator to a maximum of > 20 percent northeast of The Bahamas.
>
> (Pielke et al., 2003, p. 102)

Because of its geographical conditions, the Caribbean region is prone to such natural events of severe intensity (Charveriat, 2000). However, susceptibility to natural hazards is not uniform throughout the region. The Greater Antilles (Cuba, Haiti, and Jamaica) have been identified as the most disaster-prone group in the Caribbean and, along with Pacific islands, with unstable economies and weak political and institutional development, are heavy losers to repeated natural shocks (Pelling and Uitto, 2001). This finding is supported by Spence et al. (2005), who found that the northern Caribbean, including Jamaica, Cuba, Hispaniola, Puerto Rico, The Bahamas, Turks and Caicos Islands, and the Cayman Islands, has high inter-annual variability of hurricane occurrence with a mean strike rate of 1 per year; while the southern Caribbean experiences a much lower strike rate of 0.4 hurricane strike per year.

Nurse and Sem (2001) reported a mean annual rainfall decline across the insular Caribbean of about 250 mm since 1900, while Taylor et al. (2002) noted a marked negative trend in rainfall from the 1960s. Gamble et al. (2010), in a graphical analysis of monthly rainfall figures for St. Elizabeth Parish from 1980 to 2007, demonstrate that the frequency of longer drought events appears to have changed in the early 1990s. Considerable variability exists in the data, but a tendency toward longer and deeper drought periods since then is evident.

Mimura et al. (2007) agree that today the incidence of hurricane has increased from 1995, with 2005 being one of the most active on record (Anthes et al., 2006). The record overall thus points toward increasing temperatures, decreasing precipitation, and increasing incidence of extreme weather conditions (hurricane activity), particularly since the mid-1990s, but with significant variability both across the basin and at individual island locations. An increasingly unpredictable and, for agriculture, "risky" climate is indicated. The negative effects of neoliberal processes serve to amplify the projected impacts of these changes.

In addition, academic/scientific opinion suggests that climate change has resulted in the magnification of extreme events such as droughts and hurricanes (IPCC, 2007). For example, the years 1995–2000 experienced the highest level of North Atlantic hurricane activity in reliable record. It has been noted that "the past 6 years have seen a doubling of overall activity for the whole basin, a 2.5 fold increase in major hurricanes and a fivefold increase in hurricanes affecting the Caribbean" (Goldenberg et al., 2001, p. 474).

10.3 Climate Change and Agriculture

Domestic food production and food security have been a concern in the region for many years, but evidence of climate change and recent increases in natural disasters have brought these concerns more sharply into focus. The implications of climate change for food security and sustainable livelihoods are also causes for serious concern. In Grenada, for example, Hurricane Ivan in 2004 impacted the economy by wiping out the nutmeg industry (which accounts for a third of total GDP earnings), disrupting food supply systems, resulting in severe shortages, and leaving more than two-thirds of the population homeless (Khalid et al., 2008).

Similar negative implications exist if the climate gets drier and there is an increase in the incidence of droughts. Mehdi (2007) postulates that since the 1970s, the Caribbean region has been experiencing slightly longer and more intense drought events. In 2010, for example, the region experienced one of the worst droughts in decades (Richards, 2010), forcing countries such as Jamaica, Trinidad and Tobago, St. Lucia, and Guyana to take drastic measures to preserve existing food supplies. In Guyana, rice production and the cattle industry were particularly badly affected, with approximately 10,000 acres of rice reported to be under stress and an unestimated number of livestock death (Richards, 2010). Farmers in Jamaica, St. Vincent, and the Grenadines have reported that periods of intense drought have sparked fires on small-scale food farms, resulting in crop loss.

Increased irrigation can mitigate this pattern, but water for irrigation is already scarce and will become even less available if rainfall decreases (Brown, 2005). Myers (2006, p. 1) argues that since the "weather pattern is one where the dry season is much longer and the fact that most of our food production is rainfed, it means that the switch to irrigation and irrigation systems is not moving as fast as climate change." Dry agro-ecological zones across the region should be expected to experience the most severe impacts of these changes.

According to Tubiello and Fischer (2006, p. 1031), while some crops will benefit from climate change, the general consensus is that the

> associated impacts of high temperatures, altered patterns of precipitation and possibly increased frequency of extreme events such as drought and floods, will probably combine to depress yields and increase production risks in many world regions, widening the gap between rich and poor countries.

Developing countries such as the majority in the Caribbean with a relatively high agricultural population are more likely to be vulnerable to climate change (Parry, 1999).

Whether or not it influences the frequency, intensity, or duration of hurricanes, droughts, and other hazards, changes in climate variability will affect food production at all scales (Slingo et al., 2005; Huntingford et al., 2005; Parry et al., 2004). Within the context of a changing Caribbean climate, small farmers are, and will be, increasingly vulnerable to these external stresses and shocks (McGregor et al., 2009). Some crops will become marginalized, and will require increasing and careful use of irrigation systems as a result of decreases in water availability via increasing evapotranspiration, while others will surpass their critical temperature threshold (Slingo et al., 2005), leading to a retardation in growth and productivity.

Poncelet (1997) reviewed the evolution and accomplishments of Caribbean disaster management in the 1970s and 1980s and found that, prior to the 1970s, disaster management in the region was largely "improvised" (reactive and circumstantial). The author identified two disaster periods between 1979 and 1992, which served as impetuses for change in the disaster management approach in the region. These two periods (1979–1980 and 1988–1992) witnessed some of the worst disasters to affect the region, resulting in widespread loss of lives and destruction of properties. The failure of national governments in the region to mitigate the impact of these events exposed the flaws in the reactive disaster management system. Poncelet deduced that these changes represented two fundamental shifts in disaster management in the region—the period 1979–1980 witnessed the first major attempt to coordinate disaster response and the second period represented a shift from disaster response and preparedness to disaster mitigation.

Disasters are context-driven and embedded in the peculiar vulnerabilities of a system. It is therefore imperative that a vulnerability management approach rather than a disaster mitigation approach be taken to reduce the incidence of disasters in the region (Campbell, 2011). Coupled with structural changes resulting from trade liberalization policies, the implications of climate change for agricultural productivity in the Caribbean are far reaching.

10.4 Farming, Hazard Vulnerability, and Assessment

Vulnerability is "the degree to which an exposure unit is susceptible to harm due to exposure to a perturbation or stress, and the ability of the exposure unit to cope, recover, or fundamentally adapt" (Kasperson and Kasperson, 2001, p. 21). This definition highlights the accepted dichotomous divide of the concept into an *external* component of stresses, shocks, and perturbations to which a system is susceptible and an *internal* component that represents the resilience and capacity to cope with the external stressor (Chambers, 1989; Brooks, 2003).

Since vulnerability is a function of the external hazard and the internal coping capacity of a system, it is important to view the concept through a "livelihood lens" (Zievogel et al., 2006) to incorporate its determinants (assets, capitals, and capabilities (Bebbington, 1999; Carney, 1998) as well as to provide a holistic view of how farmers are vulnerable to different stresses (Scoones, 1998; Zievogel et al., 2006). Moser (1998) alluded to the close link between vulnerability and asset ownership of those affected. It is logical, therefore, to assume that the more assets people have, the less vulnerable they are and vice versa.

The internal components of vulnerability are significantly influenced by farmers' access to resources and by the ability of the society, community, or region to adapt to changing circumstances (Campbell, 2011). A fundamental problem for Caribbean small farmers is poor access to appropriate amounts of good agricultural land (Campbell, 2011). This is a significant factor in rural poverty and in the short-term farm management planning horizons that rural poverty generates. Small farmers must plan for the next harvest, and do not have the financial means to plan for the longer term on their own, thwarting efforts to ensure agricultural sustainability (Campbell, 2011). Government intervention can be crucial in short-term disaster relief and in longer-term adjustments, but the region's history of intervention is not encouraging.

These changes in agricultural policies and markets can in turn influence the *resilience* of communities facing climate change. As Adger (2009, p. 349) puts it, "adaptation to climate change is limited by the values, perceptions, processes and power structures within society," and it is therefore important to consider the ways in which local communities respond in the face of drought. A number of researchers have highlighted the importance of local social relations in community resilience or vulnerability. In particular, local coping mechanisms can form a "moral economy," which can guide community action and response in the face of environmental stress.

Based on the projected impacts of climate change, many scholars have agreed that the efficient and effective adaptation of those countries, communities, households, and individuals deemed to be most vulnerable should be prioritized. There is an urgent need to understand the dynamism of coping and vulnerability in the context of developing future adaptation measures. In light of this, a growing body of scholarship has emerged in recent times with the thrust toward providing an understanding of people's adaptation to extreme events and climate change (Smit and Skinner, 2002; Vincent, 2007; Adger et al., 2003). Yet only a small number of these studies focus on coping strategies as template for future adaptation options.

Vulnerability assessment has its roots in studies on food security, hazard vulnerability, and impact assessments, and can be of either descriptive or interpretive nature (Naess et al., 2006). Descriptive assessments are based on data aggregated at the national level, while interpretive assessments are locally grounded. Vulnerability assessment is location specific and uses a diverse set of methods to examine systematically the relationships between people and their environment (Hahn et al., 2009). This body of vulnerability research emerged to address issues of how communities will adapt to future climate change as well as to redress longstanding issues of inappropriate resource distribution. Luers et al. (2003, p. 255) observes that information derived from vulnerability assessment is "critical for decision makers who often must rank limited resources in the design of vulnerability-reducing interventions." However, note that vulnerability assessment is a process and not a product (Naess et al., 2006). It is a snapshot of the state of a system at a particular time, which can be altered due to the development of new circumstances (Naess et al., 2006; Campbell, 2011).

The current focus on vulnerability marks a fundamental shift away from traditional assessments, with limited analysis to the stressors toward an analysis of the system being affected, its resilience, and adaptive capacity (Luers et al., 2003). The new approach promotes vulnerability management over risk management with the belief that a reduction in vulnerability will result in a reduction in risks. The idea is that a reduction in vulnerability always results in reduced outcome risk (Sarewitz et al., 2003, p. 6).

Furthermore, vulnerability management does not rely on the nature of extreme events to determine associated risks. It is believed that extreme events are determined by the context in which they occur and therefore "the character of an extreme event is determined not simply by some set of characteristics inherent in the physical phenomena, but by the interaction of these characteristics with other systems" (Sarewitz et al., 2003, p. 4). This would lead one to conclude that identifying and characterizing vulnerability should be of paramount importance within policy-driven research agendas.

Understanding the vulnerability of a system requires an assessment of its adaptive capacity, sensitivity, and exposure. These issues have raised two fundamental challenges to vulnerability assessment, which includes its application to policy and the inclusion of local knowledge. Livelihood approaches to vulnerability assessment, where indicators are informed by household characteristics, represent a positive step toward integrating local knowledge, vulnerability assessment, and coping and adaptation strategies.

10.5 Policy Response to Hazard Vulnerability in the Caribbean

Campbell (2011) identifies five key projects and institutions through which the Caribbean region is responding to climate change vulnerability and adaptation needs. These include

1. The Caribbean Planning for Adaptation to Climate Change (CPACC) project (1997–2001), which was implemented by the World Bank to aid capacity-building efforts to adapt to climate change;
2. The Adaptation to Climate Change in the Caribbean (ACCC) project (2001–2004), which was designed as a follow-up to CPACC and was also overseen by the World Bank with the aim of furthering capacity building;
3. The Mainstreaming Adaptation to Climate Change (MACC) project, which was established during the ACCC's term with funding from the Global Environmental Facility (GEF) and sought to mainstream climate change adaptation into national development policies;
4. The Caribbean Community Climate Change Centre (CCCCC), which was established in 2004 and is based in Belmopan, Belize, and is responsible for coordinating responses to climate change in the region; and
5. The Special Pilot on Adaptation to Climate Change (SPACC) project (2007–2011) being piloted in Dominica, St. Vincent and the Grenadines, and St. Lucia to assist with the implementation of adaptation efforts.

It is clear that efforts are being made to reduce climate change vulnerability and disaster risks within the agriculture sector. These efforts are tending to occur in isolation with only limited coordination among the various stakeholders. However, the problems facing the agricultural sector are multifaceted and will require a holistic approach to fully understand the complexities involved. Even though the implications are far reaching, climate change adaptation and disaster risk reduction are only two of the many challenges facing the agriculture sector. The ramifications of climate change impacts on the livelihood of farmers, and their communities are situated within a much broader context of struggling local and regional institutions. For example, local and national government institutions continue to struggle with negative changes in social structures and international market conditions, which also fundamentally influence the adaptive capacity of farmers to cope with global change.

High-magnitude fast-onset catastrophes have dominated the attention of the disaster risk reduction community for many decades. Less attention has been paid in the Caribbean region to slow-developing geophysical hazards such as drought and climate change. Research has shown that in areas such as southern St. Elizabeth, Jamaica, droughts are often more devastating than fast-onset events such as hurricanes. For example, when asked about a specific drought that occurred after Hurricane Ivan, farmers were quick to point out that their concern was not the severity of a single drought event (they are familiar with extremes), but what seems to them like a progressive increase in their frequency and the associated depletion of their resources (Campbell, 2011). Farmers often expend considerable amounts of resources in coping with drought events but still may end up losing the battle. Therefore, disaster risk reduction policies should take into account variations in agroclimatic conditions and the experiences of farmers who experience them.

10.6 Summary

Climate change is certain to impact agriculture and food security in the Caribbean as it will across the globe. Scientific evidence already points to variability in temperature and precipitation and impacts on sea surface temperatures. There has been a growing trend of unpredictability in regional climatic patterns and an alarming trend in increased frequency and magnitude of extreme events mainly hurricanes and droughts. Staggering losses in livestock and crops have been experienced in recent times, and it is reasonable to expect that the region will see more of this in the future.

Climate change, the ubiquitous nature of hazards, and the implications for regional food security make it imperative to understand issues of coping, adaptation, and resilience among small-scale farming sectors in the Caribbean. Hurricanes and droughts do the greatest amount of damage to agriculture in the region. Hurricanes are devastating because they not only have impact through high winds but, almost always leave severe flooding in their wake. In the next chapter we discuss the impact of hurricanes and droughts on food production in the Caribbean and examine the coping and adaptation strategies of farmers. We will therefore not belabor the point here, but it is important to note that coping strategies could have disastrous long-term impacts. Smucker and Wisner (2008) studied the evolution of drought-coping strategies in Kenya, East Africa, over the past 30 years. They concluded that farmers' coping responses resulted in a reliance on measures that may actually threaten future livelihood security. This emphasizes the danger of leaving farmers entirely up to their own devices as they adjust and respond to environmental change. At the same time we should not forget the ability and success of smallholder farmers to fashion homemade solutions to a myriad of farm-level problems. Eriksen et al. (2005) analyzed the coping strategies of smallholder farmers in Kenya and Tanzania to drought and concluded that at the end of the day, it is not government interventions but rather the strategies of these farmers that are critical to managing the stresses from climate and weather.

CHAPTER 11

Hazard Mitigation: Caribbean Small-Scale Farmers' Coping and Adaptation Strategies for Hurricanes and Drought

11.1 Overview

This chapter explores the adaptation and coping strategies of Caribbean smallholder farmers against hurricanes and droughts. The vulnerability of the agriculture sector to both climate change and climate variability is well established (Parry et al., 2004; Slingo et al., 2005). The general consensus is that agricultural productivity will be reduced as a result of changes in temperature and rainfall (Slingo et al., 2005) and the increasing frequency of extreme meteorological events.

Farmers' perceptions of climate are inseparable from the rich body of local knowledge that has helped them to cognize and negotiate numerous environmental uncertainties (Campbell, 2011). Gyampoh et al. (2008, p. 3) note the important role local knowledge has played over the years in solving problems, including those related to climate change: "the ability to, and how long one can cope with or adapt to these changes depends on the resources available and local knowledge is a major resource." Farmers may not understand the science behind climate change, but because observing the environment is crucial to their livelihoods, they are well in tune with the weather and are well positioned to discern even minor deviations from what are normal conditions.

The congruence of farmers' perceptions of climate and meteorological record has been noted elsewhere (Vedwan, 2006; Gbetibouo, 2009) amidst growing interest in bridging the gap between traditional and scientific knowledge in light of the unequivocal nature of climate change (IPCC, 2007) and the desire to enable vulnerable groups to adapt. The analysis of the intersection between local and scientific knowledge offers a unique opportunity to further understand and synergize the relationships between the two areas. Gyampoh et al. (2008) proposed that, as far as climate change is concerned, local and scientific knowledge should be complementary on the basis that "models and records of precipitation mainly focus on changing amounts of precipitation with climate change, [while] knowledge of indigenous people also emphasizes changes in the regularity, length, and timing of precipitation" (Gyampoh et al., 2008, p. 5).

Because hazards pose significant threats to households, communities, and societies that are dependent on agriculture for their livelihood, the knowledge of what farmers do in response to these events can broaden adaptation options and improve resilience within the sector. Planned adaptation to future climate will be based on current individual, community, and institutional behavior, in part developed as a response to the present climate (Jones et al., 1999). Our analysis contributes to the understanding of the vulnerability, adaptive capacity, and capabilities of farmers to hydro-meteorological hazards.

The effects of meteorological events on agriculture in the Caribbean are well known. However, because farm-level adaptation is pivotal in translating climatic challenges and agricultural responses into changes in production, prices, food supply, and welfare, this discussion adds another dimension to the existing analyses by providing a detailed understanding of what farmers do to mitigate the impacts of a natural hazard event. This is important because hurricanes and droughts are ubiquities and pose a constant threat to households, communities, and societies in small island states like those of the Caribbean. The knowledge of what farmers do in response to these events can aid in the expansion process of adaptation options and the improvement of resilience within the sector as a whole.

Adaptation studies enable the identification and development of community-specific adaptive measures and practices. It has been argued that the

> aim (of community-specific adaptation studies) is not to score adaptation or measure relative vulnerabilities, nor quantify the impacts or estimated effects of assumed adaptation. Rather, the focus is to document the ways in which systems or communities experience changing conditions and the processes of decision-making in this system that may accommodate adaptations or provide means of improving adaptive capacity.
>
> (Smit and Wandel, 2006, p. 282)

11.2 Hurricanes

11.2.1 Trends in Caribbean Hurricane Activity

Hurricanes are the most prevalent meteorological hazards that occur in the Caribbean (Potter et al., 2004; Poncelet, 1997). They also have the most dramatically manifested impacts on local agriculture. Charveriat (2000) notes that the Caribbean region, due to its geographical characteristics, is prone to natural events of severe intensity with spatial variations in susceptibility. For example, the northern Caribbean (Cuba, Jamaica, Hispaniola, Puerto Rico, The Bahamas, Turks and Caicos Islands, and the Cayman Islands) have been identified as the most disaster-prone group in the Caribbean (Spence, et al., 2005) and besides the Pacific Islands, which also have unstable economies and weak political and institutional development like the Caribbean, are heavy losers to repeated natural shocks (Pelling and Uitto, 2001).

Research indicates some relationship between the nature of hurricanes and sea surface temperatures (SSTs) (Peterson et al., 2002). Lugo (2000, p. 248) posits that, while an increase in surface temperature is likely to amplify the speed of hurricanes, there is little evidence to suggest that areas affected by hurricanes would increase. Changes in SST have been attributed to the El Nino event (Chen and Taylor, 2002), which was also found to influence hurricane landfall activity in the region (Tartaglione et al., 2003). The period 1995–2007 saw an increase in hurricane activity. Between 1995 and 2000 the

> North Atlantic experienced the highest number of hurricanes in reliable record. In the last 6 years there has been a doubling of overall activity for the entire basin, a 2.5-fold increase in major hurricanes and a fivefold increase in hurricanes affecting the Caribbean.
>
> (Goldenberg et al., 2001, p. 474)

The year 2003 was especially active, with 16 tropical storms, 7 minor hurricanes, and 3 major hurricanes. Hurricane intensity also seems to have increased with an overall rise in the number of category 4 and 5 hurricanes since 1970. Emanuel (1987) supports the reporting that major storms in both the Atlantic and the Pacific Ocean since the 1970s have increased in duration and intensity by about 50 percent.

This trend is manifested in the Caribbean, which has experienced an increase in the frequency and intensity of hurricanes in recent times. Diverging interpretations exists as to whether or not SST is the driving force behind these changes or simply a feature of the natural multi-decadal cycle (Campbell, 2011). Our purpose here is not to settle this debate, but to provide an understanding of the impacts and coping strategies of farmers to hurricane events. Questions about resilience, adaptation, and vulnerability can be answered effectively only by understanding existing coping strategies and the internal characteristics of a system (Adger et al., 2005).

11.2.2 Farmers' Awareness of Hurricanes

Farmers are usually aware of hurricanes. Tropical cyclonic systems with implications for the region are tracked from the time they are spawned and their development is monitored closely. Caribbean people are updated through media reports, weather reports, and special bulletins. Yet it is not uncommon for farmers and other local people especially in deep rural areas in the larger islands to claim unawareness of a hurricane's approach. For example, Campbell and Beckford (2009) in a study on farmers in Jamaica's south coast found that two days before the storm was slated to reach Jamaica, 27 percent of farmers were not aware of the approach of Hurricane Dean. Of the 73 percent that were aware of the threat, 63 percent reported that they felt that it would strike Jamaica and thus took the threat seriously. Thirty-four percent thought the threat was not imminent as the island would not be struck. Only 36 percent of the farmers indicated they would institute measures on their farm to reduce the impact of the hurricane. The remaining 64 percent felt there was nothing they could do to save their crops except hope for the best. This is a common response of Caribbean farmers to hurricanes. They make efforts to secure the homestead and protect livestock and animals but most farmers are unaware of strategies they could take to save crops and minimize loss. But as we discuss in the next sections, some farmers did make efforts to mitigate their losses (Campbell, 2011; Campbell and Beckford, 2009).

11.2.3 Crop Protection Measures

Campbell and Beckford (2009) reported that about 60 percent of the 50 farmers modified or adjusted their farming practices in accordance with the prevailing climatic constraints. Predictably, all the farmers spent more time and resources securing their homesteads than protecting crops both before and after the hurricane. This reinforces the fact that the house is an important part of the production space of farmers (Campbell, 2011). Not only is the house a refuge for the farm family when a hurricane threatens, but it is also a place to protect farm animals and store produce. As a result, farmers will endeavor to protect their houses as a matter of priority. This represents an adaptive strategy firmly enshrined in the cultural ecology of domestic food farming systems throughout the Caribbean.

The main crop-saving strategies of farmers during the immediate period of Hurricane Dean were the protection of nurseries, (re)transplanting, crop bracing, lowering yam sticks, cutting trenches, spraying crops, as well as the harvesting and storage of produce. These strategies are primarily dependent on (i) the stage of crop growth, (ii) the type of crops grown, (iii) the terrain, (iv) the scale of production and the availability of labor, and (v) the age, health, and experience of the farmer.

11.2.3.1 Protecting Nurseries

Farmers' efforts were geared toward protecting seedlings and seed beds from the effects of high winds and torrential rainfall. They did this by using corrugated roof sheeting, wood, blocks, and grass to provide cover and barrier. Some farmers had seedlings in planting trays and were able to take them inside their house for protection. Seedbeds represent the traditional way of nurturing seedlings in the Caribbean, while trays are a modern and more efficient way to do so, and are becoming more widely used among small-scale food farmers. However, it is clear that some farmers still prefer the traditional way primarily because of the high cost of seedling trays and the treated organic matter that is required. The farmers without trays recognized the advantage enjoyed by their peers who had trays. Trays were advantageous and convenient because they were portable and could be brought in for protection from the hurricane. This was also convenient during intense rainfall, which could cause flooding and damage seedlings. Overall, efforts by farmers with natural seedling beds to protect nurseries proved effective as there were no observed or reported cases of damage to seedlings under the roof-like and tomb-like structures.

11.2.3.2 (Re)transplanting

One farmer tried the innovative strategy of (re)transplanting seedlings. He was trying this for the first time having seen a farmer in another farming community miles away try it successfully in response to an earlier hurricane. The farmer carefully lifted his two-weeks-old tomato plants, put them in boxes, sprayed the leaves with leaf fertilizer, and brought them inside his house. After Hurricane Dean passed, he replanted the tomato plants. The process is called (re)transplanting because the plants had already been planted once when transferred from the nursery to the fields. (Re)transplanting can negatively affect the growth and development of any plant as it offsets the intake of vital micronutrients, which often results in the wilting and dying of plants. Therefore, an observation of the plant's development until it reaches maturity is necessary before any reliable conclusions can be made about the effectiveness of the strategy. It will be interesting to see if this strategy spreads in this community and becomes a viable adaptation strategy. Its use by this farmer also raises interesting questions regarding a critique often expressed in the literature about local/traditional knowledge, which holds that they are localized and spatially not transferable. In this case the farmer used a method from another geographical area to seemingly good effect.

11.2.3.3 Plant Bracing and Lowering Yam Sticks

Crops such as sweet pepper and yam are difficult to protect when they reach a certain stage of growth and maturity as they should neither be covered nor uprooted. Yams simply cannot be uprooted and successfully replanted after a certain point. One sweet pepper farmer operating on a hillside decided to use crotch-sticks to brace the plants against the gradient of the hill. The farmer's experience with Hurricane Ivan was the real motivation behind this strategy. The farmer reasoned that it was the wind from Hurricane Ivan that did the most damage to his crops and so he is used the crotch-sticks to try to support the plants. The farmer did not lose all of his crops but he attributed this more to the "mercies" of Hurricane Dean than his bracing strategy.

A yam farmer did something equally simplistic to save his entire garden; he uprooted the yam sticks and placed them on the ground. According to him, it is better that the yams "fight" on the ground rather than in the air, which simply means than their chances of survival were better on the ground. Left standing, the sticks would be snapped or blown over by the strong hurricane winds. This is a very unique approach, which has not been reported before and is not a feature of the main yam-farming communities in Jamaica perhaps because of the difference in scales of production. It would be interesting to see how farmers in the main yam-growing areas would assess the feasibility of this approach as a response to hurricanes.

11.2.3.4 "Pre-wet Season" Strategies

Some farmers with fields on hillsides as well as in low-lying flood-prone areas cut or cleared trenches on their farms to channel water away from their crops. Cutting trenches to help control soil erosion is a common strategy among Jamaican farmers. These trenches are cut across slope gradient. Some farmers also use lines of rocks across the contour to help control soil erosion, which is often stimulated by torrential rainfall (Campbell and Beckford, 2009). Some farmers sprayed their crops with either leaf fertilizer or the fungicide *diathane* or a combination of both to enhance the resilience of their crops to extreme rainfall.

11.2.3.5 Harvesting and Storage

As we explained in Chapter 4, the decision-making process of farmers is complex and often seems to be counter-intuitive as it goes against what conventional wisdom might define as being logical. For example, if a hurricane threatens the island and a farmer has mature produce on his or her farm, then the logical thing to expect is that he would harvest and store them until after the event. However, before arriving at such a decision, farmers often take historical experiences into consideration. For example, based on a history of near-misses from hurricane, some farmers did not believe Dean would hit and that pre-harvesting would affect the quality of farm produce.

This manifestation of *gamblers' fallacy* (Burton et al., 1978) is common in the response of people to disasters. Gamblers' fallacy is a process of banking on the fact that extreme disasters are rare and will not reoccur in quick succession. It is similar to saying that lightening won't strike twice in the same place. With the impact of Hurricane Ivan still lingering in their minds, however, other farmers did not hesitate to harvest and store produce in preparation before Hurricane Dean could strike. This is an example of adaptation, but the different responses by farmers emphasize that adaptation often is uneven and non-uniform even over small geographical spaces.

One of the many incentives to harvest and store produce before a hurricane is that produce fetch good prices immediately following the event. In the study reported here, 25 percent of the farmers indicated that they would harvest and store produce. However, 6 percent said they would not harvest produce expecting the hurricane to behave like others in the recent past—change its course and miss the island.

11.2.4 Post-Hurricane Strategies

Adaptation is a continuous process of adjustment that seeks to provide an adequate compromise between losses and gains. The process has to make sense to the farmer within the context of his/her livelihood. In this sense, adaptation is usually "the result of individual decisions influenced by forces internal to the farm household and external forces that affect the agricultural system at large" (Smit and Skinner, 2002, p. 7).

After Hurricane Dean the efforts of farmers were mostly geared toward restoring their livelihoods to a state of normalcy. Some farmers utilized financial resources (e.g., savings) available to them to restart production and restore damaged properties. However, the majority of farmers surveyed reported limited financial assets and as a result had to depend on other forms of capital (physical, human, and social) in their restoration efforts.

Loss of crops from Hurricane Dean was significant. Thirty-six percent of the farmers reported that they had lost their entire crop. In the aftermath of the hurricane, some farmers employed a number of coping strategies. The most common included (i) post-hurricane harvesting and plant restoration, (ii) relocation of farm/plot, and (iii) scaled-down production.

11.2.4.1 Post-Hurricane Harvesting and Plant Restoration

Farmers identified marketable produce, which they harvested and tried to sell immediately. Not surprisingly, all 50 farmers sampled indicated that the price for produce was better in the immediate period after Hurricane Dean than for the period preceding the hurricane. News reports indicated that there was a 300 percent increase in the price of vegetables such as tomato, carrot, and sweet pepper in many parts of the island after the hurricane.

Farmers also tried to rehabilitate some crops. With little government relief available, they turned to a range of tried and proven traditional or customary agronomic practices to salvage crops and mitigate losses. These included weeding, molding, mulching, fertilizing and manuring, spraying, and watering. Farmers know which crops are most easily revived and which are more difficult. Sweet potato, scallion, and beetroot were identified as the easiest crops to revive, while tomato, melon, sweet pepper, and cabbage were included among the latter category.

11.2.4.2 Relocating Farm Plots

Farmers are generally hesitant to change the location of their farm plots. Chief among the factors that explain this attitude are land availability and tenure, indigenous technical knowledge—specifically knowledge of environmental conditions such as soil type, rainfall and pest and insect behavior, type of crops being cultivated, distance from homestead, access to road and water, and the availability of labor. However, after Hurricane Dean, three farmers indicated their intention to relocate farm plots. Two of the three farmers were cultivating on hillsides and planned to move to the foothill, where eroded topsoil and fertilizers and manure applied had accumulated. This is an example of the ability of farmers to adapt and be innovative and make use of micro-ecological niches.

11.2.4.3 Scaled-Down Production

A common aspect of small-scale farmer disaster management is scaling back production after an extreme event. One reason for this is the tendency for farm inputs like fertilizers to increase in price after a hurricane. More than 70 percent indicated that they would scale down production as a short term measure. Production reduction occurs both in terms of land area and number and amount of crops cultivated. In determining which crops to scale back on, farmers considered market conditions, profitability of various crops, and the complexity of growing different crops in the aftermath of the storm.

Some farmers indicated their intentions to give up farming altogether—perhaps an impulsive reaction born out of frustration—while others intended to grow a single cash crop, which increases their vulnerability. Others have found it necessary to seek off farm work during the recovery period. These changes can have serious implications for sustainable livelihoods and food security.

11.2.5 Relief Efforts and Recovery

It has been argued that in relation to hurricanes, the "period of recovery is an important factor that is given little attention and is especially important because of the nature of the annual hurricane season and possible trends as a result of anthropogenic climate change" (Ferdinand and Parker, 2005, p. 3). The period of recovery is fundamentally a function of the socioeconomic and political factors (such as effective government policies to assist affected persons after a disaster) as well as the health and well-being of the affected. The degree of damage to the environment (e.g., the erosion of valuable top soil and destruction of road networks) can also affect the recovery time of farmers. Governments can help to shorten the recovery time by working with farmers to understand their needs as well as to facilitate the implementation of community-specific policies geared toward increasing resilience. This means that robust adaptive measures must be in place. This has scarcely been the case in the Caribbean and the reported experiences of farmers are replete with statements of discontent about the disorganized and inappropriate disaster response and relief strategies. For example, after Hurricane Ivan in 2004, farmers were given a J$5,000 coupon (<US$70), redeemable only against carrot seeds supplied by a single large Jamaican agroprocessing company, and regardless of a farmer's particular cropping specialty (McGregor et al., 2009) or the level of damage and loss experienced.

The relief efforts do not ensure that the worse affected and most vulnerable farming households get assistance. After Hurricane Dean, assistance was given in the form of fertilizers and farmers received on average 1–3 bags regardless of the size of their farms. Over 84 percent of farmers reported that they were not visited by agricultural extension officers or officials after Hurricane Dean and 74 percent reported that they did not receive any assistance from the government. This suggests quite clearly that farmers were in essence left to fend for themselves as they attempted to recover. Unfortunately, this is the common experience of farmers throughout the Caribbean after natural disasters like hurricanes. Many farmers are left to pick up the pieces relying on their social network and individual ingenuity to survive. This reinforces the importance of social capital—the "features of social organization such as trust, norms and networks that can improve the efficiency of society by facilitating coordinated actions"(Putnam et al., 1993, p. 167) as a critical component of rural livelihood and adaptation. Unfortunately, many farmers do not have social networks with the capacity to help and depend more on government assistance.

While typical relief efforts in farming communities in Jamaica scarcely build resilience and, in many cases create dependency, it is difficult to overstate its importance in accelerating the recovery process in the immediate aftermath of a disaster. Recent literature has stressed the importance of assisting affected persons after a disaster not just to return to "normalcy" but to build capacity so they will be better able to respond and cope in the future (Mustafa, 2003). This is the essence of adaptability. However, poorly organized and inadequately financed relief efforts also impinge on farmers' ability to recover from such shocks and stresses (McGregor et al., 2009). The Third Assessment Report of the Inter-Governmental Panel on Climate Change emphasized the important role governments have to play in the post-hurricane adaptation process. If they are well organized and prepared, the adaptive capacity of the most exposed sectors will improve. However, "a disorganized and unprepared government will mean a lower adaptive capacity for a country" (Vincent, 2007, p. 14).

While acknowledging the importance of social capital, the literature on the subject also argue that "strong bonding ties are associated more with survival than development and are often observed in recovering from natural disaster and conflict" (Pelling and High, 2005, p. 310). Therefore, while social capital helps farmers to recover from natural disasters, it is hardly associated with an improvement in their standard of living. Put another way, strong bonding ties are helpful in coping but not so much in adaptation. However, some literature also argues that the presence of "bonds" (strong kinship network) can increase the adaptive capacity by providing economic, managerial, and psychological help (Smit and Wandel, 2006). The difficulties faced by farmers as a result of Hurricane Dean were exacerbated by a two-week rainfall caused by tropical storm Noel that resulted in severe flooding, water-logging, and destruction of crops. The continuous disruption of livelihood activities by successive extreme events amplifies the plight of these farmers who operate in a relatively marginal agricultural system and underscores a fundamental issue in the capacity of the region to achieve greater food self-sufficiency and general food security (Campbell and Beckford, 2009).

11.3 Drought

Drought is a major hazard for Caribbean farmers. Its effects are usually not as manifestly dramatic as hurricanes, but its results are no less damaging to farmers and rural economies. Droughts lead to other problems like fires, and affects not just crops but livestock as well. Coping strategies of farmers in relation to drought are developed mainly to counteract the immediate negative impacts associated with the event. Here, the term *coping strategies* is used to represent any action of the farmer geared toward mitigating the impacts of a drought episode on their farm. By definition, adaptive strategies are more sustainable than coping strategies and are better suited to deal with longer-term changes (Mcleman and Smit, 2006; Ingham et al., 2005; Cooper et al., 2008). This has led Cooper et al. (2008) to question the usefulness of coping strategies in capturing the benefits of good seasons, as coping strategies are "risk spreading" and designed to mitigate the negative impacts of poor seasons, so usually fail to exploit the positive opportunities of average and better-than-average seasons. While this may be true, the coping strategies of farmers are usually informed by many factors that combine to navigate his/her socio-spatial behavior.

Here the coping strategies of farmers are divided into four groups that explain (1) the planting methods farmers utilize to cope with droughts; (2) the different techniques used to reduce moisture loss; (3) the strategies farmers employ to cope during a drought; and (4) strategies farmers use to restore production to normalcy after a drought.

11.3.1 Planting Methods

The two most frequently used planting methods identified by farmers were the planting of "quick" crops (56 percent) and the scaling down of production during the dry season (53 percent). To farmers, a "quick" crop is one that takes less than eight to twelve weeks to mature. These crops include a variety of fruits and vegetables—lettuce, zucchini, turnip, string bean, beetroot, cauliflower and broccoli, yellow squash, watermelon, honeydew, and cantaloupe. The obvious advantage of cultivating these crops is the short time they take to mature and their marketability. The winter tourist season (December to April) coincides with the major dry season (December to March) in the area, and the cultivation of "quick" exotic crops for farmers is a dual adjustment to both events. The cultivation of several "quick" crops during the dry season is usually done with the expectation of generating enough income to "go out big" during the rainy season with their primary crop. The practice of hand-watering plants during the dry season and successfully bringing a crop to maturity is termed "catching a crop" by farmers.

While the rainy season remains important to farmers, most do not hold to clearly defined planting months. More fundamentally, the prospect and possibility of cultivating in the rainy season is determined by the nature of crop production during the dry season. Generally, farmers displayed a risk-minimizing attitude toward crop production during the dry season. More than half (53 percent) reported that they scaled down production during the dry season. Only three farmers totally avoided planting at that time of the year and, interestingly, all three were part-time farmers and perhaps could afford to do so.

Farmers also plant crops such as cassava that are more resilient to the dry conditions and have multiple uses. The cultivation of crops more tolerant to dry spells (e.g., scallion, beetroot, sweet potato, cassava) is one of the more frequently recommended strategies to adapt to the future impacts of climate change on food systems. However, the decision to cultivate a crop goes well beyond its ability to adapt to environmental conditions, and includes other socioeconomic and cultural factors that inform the decision-making process of farmers (Beckford, 2002).

11.3.2 Moisture-Loss Reduction

Moisture-loss reduction strategies are critical in areas with low rainfall and relatively high evapotranspiration rates. Successful crop production under these conditions requires effective moisture-loss reduction strategies. Some of the strategies adopted by farmers include mulching, edging or perimeter planting, drip irrigation, and managing the application of water to plants (Campbell et al., 2010). One of the most effective methods used in the Caribbean is grass mulching, which is common in areas of Jamaica (McGregor et al., 2009). Farmers cut grass after 18–20 months, when they are dry, and spread them over every inch of a farm plot. The strategy retains water and also prevents growth of weeds.

Farms also use a technique called "edging" to reduce soil moisture loss. Edging is the cultivation of taller, more bushy plants (including guinea grass) around the perimeter of a farm plot to interrupt the flow of wind over the garden. Cassava, corn (maize), eggplant (aubergine), and gungupeas (pigeon peas) are the most common crops used to "edge" gardens. Some farmers also use the drip irrigation technique to reduce water loss and use water more efficiently. Next to mulching, farmers regard edging and drip irrigation as the most effective water-loss reduction strategies. However, a drip irrigation system is expensive to set up and is not an option for most farmers. Only two farmers in this sample had drip irrigation installed on their farms.

11.3.3 During Droughts

Coping strategies during a drought event are centered on measures to supply or better retain soil moisture. However, the most common practice during a drought is to sacrifice a portion of the crops under cultivation although this will be considered only as a last resort. Other strategies employed during a drought include thicker mulching (i.e., applying a second layer of guinea grass), buying water, sharing water, and spraying plants with leaf fertilizer.

Elsewhere in the Caribbean, other coping strategies have been observed. In the dry limestone island of Antigua, for example, farmers report that to avoid high evapotranspiration, they water plants late in the evening after sunset. Plants are therefore able to make efficient use of water. Farmers in Dominica also report the same practice with some suggesting that early morning watering before the sunrise is just as effective.

11.3.4 Drought Recovery

Scaling down production to a smaller, more manageable level and planting fewer and different crops is a common response to drought (Campbell et al., 2010). Land taken out of production is usually left uncultivated. Some farmers temporarily exit farming, others focus on a single cash crop (which increases their vulnerability), while other farmers engage in a variety of casual off-farm work for pay. Off-farm work is an important disaster recovery strategy for farmers (Campbell et al., 2010). In the case of drought, the impacts are not as sudden as hurricanes. Farmers can see the disaster unfolding before their eyes and are able to plan better. Still these changes have serious implications for sustainable livelihoods and food security. The number of farmers who can no longer afford the cost of farming is increasing and likewise a state of dependence, insecurity, and vulnerability to extreme weather events such as droughts is also increasing.

11.4 Summary

There is an urgent need to understand the dynamism of coping and vulnerability in the context of developing future adaptation strategies (Campbell and Beckford, 2009). A growing body of scholarship is emerging in response to this need geared toward providing an understanding of peoples' adaptation to extreme events and climate change (Smit and Skinner, 2002; Vincent, 2007; Adger et al., 2003). However, only a small part of this scholarship focuses on coping strategies as a blueprint for future adaptation options (Campbell and Beckford, 2009).

A coping strategy may be defined as a temporary action, reaction, or response to an event or episode. This is different from adaptation that is best seen as a sustainable response in a social, ecological, or economic system that goes beyond simply coping or returning to normalcy to building resiliency in preparation for future events. In the context of agriculture, "coping strategies are risk spreading in nature and designed to mitigate the negative impacts of poor seasons and usually fail to exploit the positive opportunities of average and better than average seasons. In addition, farmers often overestimate the frequency of negative impacts of climate variability, and underestimate the positive opportunities" (Cooper et al., 2008). They argue that in order to enhance the adaptive capacity of agricultural communities and stakeholders, their ability to cope with challenges and opportunities of current climate variability must first be understood.

Of course, coping strategies are not always easily identifiable given that the diversification of economic activities is a characteristic livelihood feature in rural areas (Lambert, 1994) and given that the coping strategies of some farmers may be regular livelihood activities of others. Lambert (1994) agreed that coping strategies can enhance the resilience of people to external shocks but cautioned that these strategies enabled people to cope with crises and that we should be mindful when reinforcing them not to perpetuate a cycle of coping and subsistence. This is a valid perspective, but it should be pointed out that what might start out as a coping strategy could over time turn out to be an adaptation.

The coping strategies of Caribbean small-scale farmers are buffers against short-term shocks and stresses and often exist alongside more-long-term adaptive strategies. Coping and adaptive capacity among farmers is fundamentally a function of their livelihood assets (Chambers, 1989). Stability and diversity in resource base make them better equipped to respond to extreme climatic events (Cooper et al., 2008).

As we established a number of times in this volume, meteorological hazards are ubiquitous to the Caribbean, and small-scale food farmers are well acquainted with their impacts. We argue that for the region to achieve a greater degree of food security, the small-scale farming sector must improve in adaptive capacity to hurricanes. The damages and crop losses due to hurricanes have been staggering. There is often very little relief as in recent times, extreme events have seem to occur in series or waves. It is common for an intense hurricane season to be followed by an extended drought, which is then followed by torrential or intensive rainfall leading to flooding or landslides. Farmers are often hit again while they are still recovering from an event. For example, many farmers had still not recovered from Hurricane Ivan three years earlier when Hurricane Dean had hit. Adaptation strategies need to be more sustainable, which according to Adger et al. (2005) is dependent upon the heterogeneity of adaptive capacity across different stakeholders. They must also be robust and flexible (Adger et al., 2005) features, which we argue are enhanced when grounded in local knowledge about the agro-ecological system.

Our research underscored the urgent need for more state institutional support for the agricultural sector in general but for the small-scale food producing sector more specifically. Governments in the Caribbean have tended to invest in disaster planning systems, which focus on mitigating loss of lives. We argue very strongly that strategies to protect rural livelihoods must also be incorporated. Given the importance of agriculture to rural livelihoods, the Caribbean economies, and regional food security, steps should be taken to moving beyond helping farmers to cope in the aftermath of a shock event to building adaptive capacities in rural farming communities. Agricultural extension services must be expanded and intensified to include hazard mitigation education for farmers and rural dwellers.

CHAPTER 12

Sustainable Agriculture and Domestic Food Production: Adaptation and Experimentation on Small-Scale Food Farms in the Caribbean

12.1 Overview

Research speaks to the tendency to see small-scale farmers in the tropics as bounded by tradition, conservative, resistant to change, lacking in innovativeness, and even lazy both in official pronouncements and in academic discourse (Richards, 1985; Chambers, 1989; Rhoades, 1989). One reason for this is the chasm in communication between farmers, on one hand, and researchers unfamiliar with the rationale for traditional agricultural practices, on the other, a situation compounded by the lack of historical records. Rhoades (1989) explains this situation saying that farmers as a rule do not document their activities and accomplishments and do not write about their discoveries and innovations. Moreover, the relative successes of the Green Revolution technology, encouraging the view that Western science and technology can solve all problems, has not only served to marginalize the ongoing efforts and diverse productions of small-scale farmers, but has also led to numerous instances of counterproductive interventions in a variety of agricultural and food-related problems in developing economies (Jones and Carswell, 2004). Carswell (2004) argues that agricultural narratives developed in colonial discourses continue to implicitly blame small farmers for poor cultivation techniques and environmental degradation, and need to be challenged in the face of mounting contradictory evidence. In the Caribbean, for example, the entire production of domestic food crops is attributed to small-scale farmers, which underscores their significance in regional food security.

But their experimental innovations have not generally been considered for on-farm research trials. Yet there is abundant evidence that small-scale farmers are adaptive and experimental problem-solvers and experts at devising innovative sustainable practices. Caribbean small-scale farmers' experimentation, adaptation, and innovation go directly to their resilience, coping strategies, and ability to survive in a harsh agro-ecological environment characterized by deeply entrenched institutional biases against the small-scale farming sector. Small-scale food producers must depend on their individual and collective ingenuity to survive and prosper, and farm- and village-level experimentation and innovation are fundamental in this regard.

This chapter highlights some examples of farm- and village-level experimentation and adaptation among small-scale food farmers. The examples include strategies to conserve scarce inputs, adapt to moisture deficiency, adapt to scarcity of land, take advantage of unique ecological niches, and provide affordable labor, among others.

12.2 Introduction

Successful adaptations and experimentations of small-scale farmers have received very little consideration in on-farm research trials perhaps because those in the traditional sector are typically seen as recipients and not creators of technical knowledge and sustainable practices (Beckford et al., 2007). However, there is considerable evidence that smallholder farmers are adaptive, problem-solve through experimentation, and have been successful in fashioning innovative survival strategies. According to Beckford et al. (2007), the literature on the topic is rich in examples from Africa, Asia, and Latin America, but until relatively recent times, there was very little documentation of the farm- and village-level adaptation and experimentation, which characterizes small-scale farming and food production in the Caribbean.

In the 1960s, insights into traditional farming cultures in the tropics proliferated, thanks to the contribution of scholars from many different academic disciplines including geography, sociology, anthropology, and agricultural economics whose research began to elucidate the circumstances and in situ practices of the smallholder farmers (see Blaut, 1959a; Mintz, 1985; Innis, 1997). Research by the likes of Hills and Iton (1983), Chambers (1989), and Innis (1997) reinforced the view that the longstanding pejorative views about the abilities and contributions of smallholder farmers were unearned. More recent research has unearthed more hard evidence of the ability of small-scale farmers to adapt, innovate, and problem-solve (Davis-Morrison and Barker, 1997; Davis-Morrison, 1998; Beckford, 2002; Bailey, 2003; Beckford and Barker, 2003; Beckford et al., 2007; Campbell and Beckford,

2009; Campbell, 2011). Expertise from village- and farm-level experimentation is derived fundamentally from lived experience in and knowledge of the environment. Scott (1977, 1987) points out that this was often spurred by socioeconomic goals and motives that were not always tied to profit maximization. Research from many parts of the tropical developing world confirm that small-scale farming systems are adaptive to environmental contexts in ways that are sustainable and contribute to food security (Chambers, 1983, 1989; Richards, 1985; Cashman, 1989; Rajasekaran, 1992, 1993) by providing livelihoods and daily subsistence for more than 1 billion people across the world (Scott, 1977). Still it could be argued that they have not received full legitimacy outside of the academic literature as policy-makers continue to marginalize their contributions and potential.

12.3 Examples of Farm-and Village-Level Adaptations and Experimentations among Small-Scale Farmers

Rhoades (1989) suggests that adaptation, experimentation, and innovation may be driven by farmers' need to satisfy their curiosity, solve problems, or adapt to new technologies. Whatever their motive, adaptation, experimentation, and innovation are characteristics of virtually all tropical smallholder farming systems. These traditional farming communities are often rich in ecological knowledge about their immediate surroundings (Wigley, 1988) and although their skills and abilities are not fully recognized by policy-makers, they are experts in survival (Bunch, 1989; Chambers, 2001).

12.3.1 Sustainable Resource Use

Farmer adaptation and experimentation is often seen in sustainable resource use. A good example of this comes from the cultivation of yams in the main yam-growing areas of Jamaica. Here farmers have resorted to the reuse of a scarce agricultural input to yam cultivation. The yam plant is a climber and is planted in individual mounds of soil called yam hills, each of which has a supporting pole from a sapling called a yam stick. The yam vines climb the pole creating an aerial biomass that, through the exposure of green leaves to sunlight, enhances the growth of the yam tuber. In some of the islands in the eastern Caribbean such as Barbados, yams are planted without stakes but in Jamaica, where yam has become an important export crop, the use of yam sticks is a ubiquitous part of the cultural ecology of yam cultivation.

A yam stick is typically 3–4 meters tall, and ideally 17–25 centimeters thick. In the past farmers used yam sticks from premier hardwoods, which could last for well over 30 years. These sticks were once easily sourced from

nearby forests and woodlands. However, increased yam cultivation intensified uncontrolled harvesting of yam sticks from immediate local sources
resulting in the depletion of local resources and the development of an informal commercial trade in yam sticks with farmers forced to purchase sticks
from dealers who truck sticks from far afield into yam-growing communities.
Several related issues have combined to create a problem in obtaining good-
quality sticks in abundant quantities at affordable prices (Beckford, 2000;
Beckford and Barker, 2001).

Yam farmers in the principal yam-growing areas of the country first
responded to this problem by reusing or recycling sticks to reduce the need to
purchase new sticks annually. Without recycling it is estimated that farmers
would be required to replace over 80 percent of their yam sticks each year
(Beckford and Barker, 2003). Because of the generally poor quality of sticks,
which is manifested in their short life spans, those that have been used in the
preceding year are often too weak to support a yam plant on theirs own the
next year. Multiple staking of yams has therefore become a feature of yam
cultivation, with two or three sticks commonly supporting a single yam plant
and as many as six being observed. Multiple staking has reduced the demand
for new sticks and this has obvious environmental benefits. The problem
in obtaining adequate supplies of good and affordable yam sticks has also
spurred farm-level trials with alternative species such as bamboo, which grows
wild, can be harvested freely, and is less expensive than hardwoods. The use of
bamboo as an adaptive response to wood shortage highlights another example
of farmers' adaptation and experimentation. Bamboo can become infested by
insects, causing it to rot quickly and collapse. Also, as a species of grass, bamboo is very invasive and can cause problems on a farm. Farmers have, however
devised ways of adapting bamboo to their needs. For example, they do not
harvest sticks during the full moon phase as insects tend to lay their eggs in
plant sap, which increases the likelihood of infestation, and they remove the
lower node, or as farmers say the "eye," from bamboo sticks and dry them
out a bit before using them to reduce the likelihood of them growing.

In some parts of the eastern Caribbean, such as Barbados and parts of
St. Vincent and the Grenadines, yam sticks are not used and the yam is
allowed to trail on the ground. Jamaican yam farmers in a recent study felt
that yams could be successfully grown without sticks and that they would be
willing to try this system.

12.3.2 Response to Moisture Deficiency

The reality of limited access to resources and services including research has
forced many resource-poor small-scale farmers to draw on their experimental
knowledge to devise adaptive responses to agricultural problems. In chapters

10 and 11, respectively, we saw how Caribbean small-scale food farmers perceive climate change and develop coping and adaptation strategies to hurricane and drought. In this chapter, we will start by discussing a specific adaptation to moisture deficiency, which has resulted in successful sustainable food production among fruit and vegetable growers in the south St. Elizabeth Parish in Jamaica. The area receives low annual rainfall, and droughts—which are increasingly becoming unseasonal—are common. Irrigational infrastructure is almost non-existent, and farmers have developed creative solutions for survival. Some have developed drip irrigation systems connected to their domestic water supply—a very expensive option. Some have put in rudimentary sprinkler systems also connected to the domestic water supply. Farmers in the area are obliged from time to time to purchase water, which is trucked into the area. The water is stored in rain barrels strategically located on a farm (see Figure 2.1). Farmers then use smaller containers and watering cans to apply water to individual plants (see Figure 2.2). This is a tedious and backbreaking task, but farmers recognize that it is a very efficient and sustainable practice that conserves a scarce valuable resource and prevents overwatering (Campbell, 2011; McGregor et al., 2009; Beckford et al., 2007; Beckford and Bailey, 2009). Purchasing water is very expensive for most farmers, but if this option was not available, many crops would not survive during drought episodes. Farmers are strategic in terms of their timing of when water is bought and which plants it is used on. Like the recycling of yam sticks discussed earlier, homemade irrigation systems and manual watering of individual plants provide proof that smallholder farmers are efficient managers of scarce resources and engage in sustainable agriculture (Beckford and Bailey, 2009).

A key moisture-retention practice observed in this area and not common in other small-scale farming systems in Jamaica or the wider Caribbean is the use of a ground cover of dried grass mulch in vegetable fields (Figure 12.1). The grass mulch retains moisture, keeps the ground cool, and keeps down weed and is a ubiquitous feature of farming in the area. Other benefits of mulching identified by farmers include enhancing germination, adding to soil fertility through grass residue, and protecting the land from rain splash and overland flow of water (McGregor et al., 2009). This practice is fully integrated into the cropping patterns in the area, with most farmers reserving some portion of their holdings to grow guinea grass, which is the species of choice for this practice. Farm land that is "resting" (that is, fallow) is typically allowed to sport a vegetation succession of grass. While all crops are mulched, farmers report that crops that produce their food above ground—melons, tomatoes, and cucumbers, for example—benefit more from mulching than root crops and tubers such as sweet potato and cassava (McGregor et al., 2009).

Figure 12.1   Grass mulching for moisture retention

An interesting practice within the general culture of grass mulching, which reinforces that local farmers experiment and devise sustainable agronomic practices, is the experimentation with a different species of grass for mulching. This occurs although guinea grass (*Parnicum maximum*) has been used and favored for generations. But some farmers now say that a species of grass known locally as cow grass (*Paspalum conjugatum*) is better for mulching (Beckford and Bailey, 2009). Farmers say that this species decomposes more slowly and so lasts longer. The drawback to this species, which was introduced in the 1970s as a pasture grass, is that it is a creeper and is hard to harvest. Though it is very invasive, some farmers no longer see it as a nuisance to be eradicated. Campbell (2011) reports that some farmers are also using other types of grass such as Baccarie, Pongola, and Seymour grass as the favored guinea grass becomes harder and more expensive to procure.

Farmers will vary the thickness of the mulch based on environmental conditions. A thicker layer is applied during droughts and in the summer than in the rainy season (Campbell, 2011). The thickness of the mulch layer also varies depending on the crop with melons, tomatoes, and carrots identified as crops that need thicker mulch. In certain communities with cooler microclimate and lower rates of evapotranspiration, smaller quantities of mulch can be applied especially in the rainy season.

12.3.3 Farm Fragmentation

Farm fragmentation is a major characteristic of small-scale farming systems in the Caribbean. Farm fragmentation is the holding or operation of farm land in more than one parcel and describes the dispersal of farm plots over a wide area, resulting in noncontiguous plots (Beckford et al., 2009). It is often seen as an unsustainable practice that adversely impacts farm efficiency and productivity. It is considered inefficient as valuable time could be spent traveling between plots, and depending on the number of plots operated or to be visited, not enough time is spent on each. The practice is also deemed to be wasteful of land and added inputs, and therefore is a problem to be rectified. Our studies, however, lead us to argue that the practice should be seen as an example of rational and sustainable resource use in certain contexts.

Fragmentation is typically a response to the inability of farmers to procure land or user rights to land in adequate amounts in a single parcel. However, we have also observed that the practice is sometimes a deliberate strategy to rationally allocate resources in response to spatial variations in environmental conditions, to spread risks, and to diversify farmers' operations. Fragmentation—which from our observations in the Caribbean is practiced by far more male than female farmers—therefore occurs as farmers try to exploit different and specific ecological niches and take advantage of unique favorable soil and environmental conditions and microclimates. In Jamaica, for example, male yam farmers cultivate plots inside the Cockpit Country rain forest to make use of the microenvironment of rich, free draining bauxitic soils that are uniquely suitable for yam cultivation. Farmers also tend to seek land that may be far from home, where they have special benefits for the cash crops that represent the mainstay of their livelihood. Yam cultivation in Cockpit Country, Jamaica, is a perfect example of this. It has also been observed that crops that require more attention and field maintenance are planted on plots closer to home. Farmers in Dominica and Antigua report that they generally planted vegetables and fresh garden fruits such as melons closer to home because they require more care. On the other hand, some farmers from Guyana report that they have planted vegetables on plots far from their homes because of greater access to rivers for irrigation purposes. This underscores the multitude of rational reasons for patterns of fragmentation among smallholder farmers in the Caribbean. Another pattern of fragmentation seen in the Caribbean is the relationship between farm size and the prevalence of fragmentation. Larger farms tend to correspond more frequently with a higher number of farm plots, while conversely, smaller farm size shows fewer farm plots (Campbell, 2011).

12.3.4   Customary Labor Solutions

The key production activities of traditional small-scale farming systems in the Caribbean—land preparation, planting, fertilizing, weeding, and harvesting—require large inputs of manual labor, which by Caribbean standards is quite expensive. Resource-poor farmers who see only modest returns on their inputs and efforts have devised creative ways of procuring labor through customary and traditional arrangements, which typically do not involve any monetary exchange. Two common informal labor arrangements are still practiced in farm communities across Jamaica. The first is called "day-for-day" and is operated by a farmer providing a day's labor on the farm of another farmer with the unwritten understanding that this will be reciprocated at some mutually agreed time in the future. Usually this labor arrangement is utilized when farmers have pressing work on their farm, which is beyond the capacity of the farmer and his family to adequately attend to in a timely way. The other arrangement is called "partnership" and functions like a labor cooperative. In this arrangement, a group of three to six farmers form a collaborative work group that alternates between working on the farm of different members on the basis of a set of mutually understood guidelines. In both cases the arrangement is such that the work performed is roughly equal. No money changes hands, but the host farmer is expected to provide food—"first one" (breakfast) and lunch—cigarettes, and so on. Both male and female farmers can be involved though men are the overwhelming participants. These reciprocal labor arrangements enable critical farm chores to be completed in a cost-effective way, which depends on but also builds and strengthens social capital and community networks. This is critical to food production and sustainable rural livelihoods.

Other examples of creative labor procurement have also been observed. In Jamaica, Guyana, and St. Vincent and the Grenadines, cases have been reported where well-to-do farmers with excess land provide user rights to landless farmers who in return provide labor on their farms under pre-arranged conditions. Sharecropping arrangements are also practiced in Jamaica, St. Lucia, Antigua, and Trinidad. Unfortunately, these customary arrangements for labor are not as prevalent as they once were and farmers are increasingly forced to pay daily wages for labor. As we might expect, female farmers are particularly vulnerable.

12.3.5 Modified Minisett Yam Cultivation

Beckford and Barker (2007) documented a classic example of small-scale farmer experimentation within the context of sustainable agriculture. They recount their experience with a Rastafarian farmer on the remote northern boundaries of Cockpit Country, Jamaica, who, contrary to the general negative views among Jamaican yam farmers, prefers minsetts to the traditional yam stick method of cultivation and claims better results with certain yam varieties. After his early introduction to minisett through the Ministry of Agriculture extension officers, he conducted his own informal experiments and independently modified aspects of the system to satisfy his needs. The promotion of minisett champions its suitability for the cultivation of yellow yam varieties. This farmer uses minisett primarily with sweet yam and has also successfully tried it with other varieties such as white yam, renta yam, and St. Vincent yam. Contrary to agricultural officials, he is adamant that minisett is not suitable for yellow yams. He reported that he produced sweet yams weighing 4–5 pounds, which is larger than the average minisett for yellow yam varieties touted by the experts. This is itself worthy of note since a major factor in the rejection of minisett by yam farmers is the small-size tubers produced, which farmers argue are less marketable than the yams from the traditional yam stick method.

This farmer through independent farm-level experimentation successfully modified the original minisett package in a number of ways, which hold lessons for agricultural authorities. First, he tried minisett cuttings in sawdust in a nursery as suggested by the Ministry of Agriculture, but changed to normal field soil when he noticed the cuttings tend to rot in sawdust. Seedlings are transported to prepared beds where he uses grass mulch combined with *bush and bramble* (an assortment of twigs, small sticks, and branches) instead of the expensive plastic mulch, which some farmers argue would be bad for yams in very hot humid conditions. While the Ministry of Agriculture recommends continuous mounds for growing yam minisetts, this farmer integrated individual mounds into his system. He sells his entire minisett output and uses traditionally produced yams as cultivars for the next crop. This farmer through his experimentation has become an opinion leader in the community. He has influenced other farmers to try minisett and had a group of four others involved whom he provided with guidance to use his system (Beckford and Barker, 2007).

This example illustrates how individual farmers and groups of farmers engage in sustainable on-farm experimentation that can create practical solutions not intended by agricultural planners and scientists.

12.3.6 Sustainable Agriculture Through Organic Farming

Organic farming has become a buzzword and is often defined as being synonymous with sustainable agriculture. Many smallholder farmers in the Caribbean engage in agriculture devoid of the use of chemical inputs without referring to it as organic farming. To be fair, organic farming in the Caribbean is a feature mainly of subsistence cultivation. The backyard kitchen garden, which is a feature of rural household livelihood, is a typical site for crop growth where chemical fertilizers and industrial pest management chemicals are not used. These small-scale gardens utilize livestock and poultry manure from cattle, goats, donkeys, pigs, and chickens that are reared at home. They also utilize kitchen waste, which serves as green mulch for the garden. In some communities in Jamaica and the Lesser Antilles, rat bat manure is harvested from caves and old abandoned buildings. Areas of kitchen gardens occupied mainly by large tree plants also benefit from animals that are tethered there for grazing. Because of their make-up, kitchen gardens are naturally detritus rich and ecologically balanced. As we will discuss in Chapter 13, kitchen gardens play a monumental role in food security and rural livelihoods in the Caribbean.

Agriculture with significant commercial orientation is less organic than subsistence farming. This is especially so for crops that are sold in local produce markets and even more true for food crops that are sold internally and compete with produce from other regions and countries. For example, banana growers in Jamaica, St. Vincent and the Grenadines argue that they have to use chemical fertilizers in order to be able to compete with banana growers from Central America. This issue is even more serious in relation to the use of chemicals for pest and disease management. Farmers argue that in order to be competitive, their produce must be first rate. Blemishes and deformities are not acceptable. They are therefore forced to use copious amounts of chemical inputs. Vegetable farmers in the region also make the same point. This shows how globalization and market forces impact agricultural sustainability.

Still in terms of the use of industrial fertilizers and chemicals, Caribbean small-scale farming is relatively sustainable. Because these farmers generally have a low resource base, most of them cannot afford those expensive inputs. There are many examples of small-scale farmers using traditional practices to manage pests and diseases. Crop rotation, mixed farming using symbiotic crop combinations, and fallows are used to good effect by many Caribbean farmers.

Despite these sustainable practices, there are problems with pesticide use among small-scale food farmers in the Caribbean. Semple et al. (2005) discussed the marked change in attitudes of small-scale farmers in the region toward the use of agrochemicals. He cites Edwards (1961), who reported the reluctance of Jamaican farmers to use inorganic fertilizers to improve production, choosing instead to rely on traditional methods of crop combinations, rotation, and organic fertilizers. By the 1980s there was a remarkable shift in attitude with researchers reporting extensive use of agrochemicals among small-scale farmers (Gooding, 1980; Grossman, 1992, cited by Semple et al., 2005). Grossman reported that dependence on pesticides in St. Vincent had increased considerably among banana growers. He attributed this to the rapid expansion of the banana export industry, but noted that applications on food crops cultivated for the domestic and regional market had also increased. Semple, et al (2005) argued that rapid urbanization and increased demand for more and better-quality food was putting pressure on farmers to use fertilizers, pesticides, and herbicides to enhance productivity. Grossman (1998) added that the increased participation of Caribbean small-scale farmers in the global food production system increasingly requires them to use agrochemicals in order to meet the quality required by international buyers. This point is supported by research from Grenada and Jamaica, where farmers linked their high use of agrochemicals to the demands of the market—foreign and local (Semple, et al., 2005). Grossman and Semple et al. are correct but micro-economic factors should be considered as well. For example, small-scale farmers from Jamaica, Dominica, and St. Lucia report that their use of herbicides had increased because it was cheaper for clearing land than hiring manual labor.

Research also shows a link between the use of agrochemicals and the importance of crops in the cropping system (Semple et al., 2005). Farmers involved in commercial farming for the local or export market were most likely to be using agrochemicals. It was observed that the more profitable the crop, the higher the level of agrochemical application. In Grenada, short-term high-value crops such as vegetables saw more application than long-term tree crops. In Jamaica, crops grown for the export market received more chemical treatment but root crops grown for sale in the local produce markets were also experiencing increased use.

Interestingly, farmers in both Grenada and Jamaica conceded that the use of agrochemicals was harmful to the environment and humans. They noted that organic manure and biological and integrated pest/crop management practices were more sustainable but took longer to show results (Semple et al., 2005).

The increased use of agrochemicals in food production raises some serious concerns. The first one that we will look at is the proper or improper use of these chemicals. Grossman's research in St. Vincent (1992, 1998) and Semple et al.'s work in Grenada and Jamaica (2005) express alarm about the limited use of protective gear generally and the absence of protective gear and measures in many cases. It was found that many farmers wore rubber boots but very few wore coveralls, face masks, eye protective gear, and respirators (Semple, et al., 2005). Grossman (1992) actually found a village in St. Vincent where farmers deliberately applied agrochemical barefooted as this made them more sure-footed on slippery slopes. The inappropriate use of agrochemicals is often attributed to a lack of education among small-scale farmers, but research evidence would suggest that there is more to it than that. The main explanations offered by farmers for limited use of protective clothing and gear is the high cost of such gear and devices and the discomfort associated with wearing them (Semple et al., 2005).

Of critical importance too is the perception of farmers regarding the level of risk involved in using these agrochemicals. Both Grossman (1992) and Semple et al. (2005) found that farmers did not associate a high level of risk with the agrochemicals they used. Grossman reported that farmers associated toxicity with the intensity of odor from the chemicals. This raises serious questions about the knowledge and awareness of farmers about agrochemical use.

There are also concerns about farmers' ability and willingness to read and follow the instructions for using specific products. Mixing rates above the manufacturers' recommendations are not uncommon sometimes deliberately to accelerate the effect of the products. It should be pointed out that commodity associations and agricultural authorities do provide some training on the proper use of agrochemicals to farmers. For example, members of the associations involved in banana export in St. Vincent and Jamaica and cocoa export in Grenada received this training.

Another major concern relates to the disposal of chemical containers and cleaning of equipment. Some farmers do dispose of containers properly by burning or burying them (Grossman, 1992, 1998; Semple et al., 2005). There are, however, many instances of improper disposal, storage, and cleaning. Containers may be reused—for storing water, chemicals, kerosene oil, and so on. Grossman reports that they could also end up as flower pots. Often containers like bags and bottles are simply thrown away in bushes adjacent to the farm. Improper disposal has implications for livestock and poultry as well as water resources.

In terms of storage, there are as many examples of improper storage as there are of proper storage. Chemicals improperly stored in kitchens or in houses have been mistaken for cooking ingredients and ingested, leading to illness and death. Semple et al. (2005) reported that 36 percent of farmers in his Jamaica study had become ill or died from chemical poisoning. Farmers identified the main symptoms of poisoning as itching of the skin, burning of the eyes and nose, vomiting, and nausea.

12.4 Summary

There are many examples of sustainable agriculture among small-scale food farmers in the Caribbean. These are in many cases linked to farm- and village-level experimentations by individuals or groups of farmers. Such experimentation and adaptations have resulted in efficient food production and contributed to household food security and rural livelihoods for thousands of farm families across the Caribbean.

But there are examples of unsustainable farming in the region as well. No more is this evident than in the changing attitudes to agrochemical use in the past four and a half decades or so. Not all farmers misuse chemicals. In fact, many are quite scrupulous in their use of agrochemicals. But problems are clearly present. Lack of protective clothing and gear, incorrect dosage, improper storage, disposal and cleaning of containers, impact on human health and life, implications for livestock and poultry, and impacts on water resources are pressing issues.

In light of the huge implications for the misuse of agrochemicals, steps must be taken to enhance sustainability in this aspect of farming. Education of farmers through resources and on-farm adaptive field demonstrations and experimentations is important. Also important is stricter guidelines and regulations about chemical importation and use.

CHAPTER 13

The Role and Value of Local/Traditional Knowledge in Caribbean Small-Scale Food Farming Systems

13.1 Overview

Local knowledge may be defined as dynamic and complex bodies of know-how, practices, and skills that are developed and sustained by peoples/communities with shared histories and experiences (Beckford and Barker, 2007, p. 118). It provides a framework for decision-making in a plethora of social, economic, and environmental activities and livelihoods among rural peoples. Local knowledge has played an active role in the lives of rural communities in virtually every part of the world. In the Caribbean, traditional cropping systems based on local and informal knowledge have been practiced since the earliest times when people inhabited the islands. This continued during the slave era, and today local knowledge and practices are still fundamental to contemporary small-scale farming in the region. The legitimacy, relevance, and importance of small-scale farmers' knowledge of their agro-ecological environment and the implications for food security in the region are undeniable.

Local traditional farming knowledge in the Caribbean is a valuable cultural capital and a tangible resource. Traditional knowledge is driven by, and is a response to, the everyday demands of farming in a challenging, economic, political, environmental, and enigmatic sociocultural context. Knowledge is constructed in real contexts, typically characterized by economic and environmental uncertainties and risks. This knowledge then becomes a vital component in coping and adaptation strategies. Part of the epistemic foundation of traditional knowledge is related to the asymmetrical power structures within which it evolves and exists and to issues of independence, self-reliance, and problem-solving, which are fundamental to successful farming and rural development.

The individual and collective ingenuity of small-scale farmers is fundamental to the survival of the domestic food production, rural development, and food security in the Caribbean. In the context of the sociocultural, economic, political, environmental, and global realities of small-scale food farming in the region, local traditional knowledge is indispensable. Small-scale farmers traditionally operate in a context where they have had to fend for themselves. Local solutions based on traditional knowledge and wisdom, and farm-level experimentation, are *sine qua non* in the ability of these farmers to survive and prosper. In this context, traditional agro-ecological knowledge is a critical safety valve that has evolved and developed out of necessity. The coping strategies and resilience, which are characteristic of the tropical small-scale farming systems in the Caribbean, are based in this intergenerational knowledge.

Scoones and Thompson (1994) posit that local knowledge facilitates dynamic information systems, which are important aspects of decision-making. Developed over time, local knowledge is influenced by social, economic, cultural, and environmental factors (Beckford and Barker, 2007). It is intergenerational and adapted to culture (the Netherlands Organization for International Cooperation in Higher Education, 2003).

In the Caribbean, local farming knowledge is shaped by on-going experimentation. It is developed outside the formal education system and is embedded in and bounded by culture and tradition (Beckford and Barker, 2007). It is a critical factor in decisions about food security, human and animal health, education, and resource management. It is characterized by several interrelated factors:

> Firstly it is localized in nature and more often than not traditional in context. Secondly it is unique to specific environmental and cultural conditions. Thirdly is knowledge constructed in informal settings, it is orally transmitted and it is rarely documented. It is dynamic, adaptive and holistic in nature and is a significant part of the way of life of rural peoples.
>
> (Beckford and Barker, 2007, p. 118)

Studies and literature about traditional knowledge in small-scale food production systems are many. The majority of these studies are from Africa, Asia, and Latin America, and there is a paucity of studies for the Caribbean (Beckford et al., 2007). In this chapter we discuss the role of local knowledge in small-scale domestic food production systems in the Caribbean. We discuss a few concrete examples of local knowledge and how their application impacts food production and food security in the Caribbean, but we are mindful of avoiding to focus too much on descriptive accounts of numerous cases of location-specific indigenous knowledge—an approach found all too often in the literature. We examine the literature on local knowledge exploring the bipolar tensions with Western scientific knowledge and some of the problems with traditional knowledge.

13.2 The Importance of Local Knowledge in Tropical Small-Scale Farming Systems

Barker and Beckford (2006) discussed how traditional knowledge has progressed through several etymological transformations, such as folk science and village science (Howes, 1979; Howes and Chambers, 1979; Richards, 1979, 1985), ethnoscience (Brokenshaw et al., 1980), and rural people's knowledge (RPK) (Schoones and Thompson, 1994). Indigenous knowledge is also in current use as is traditional knowledge and is especially popular in the literature about Aboriginal groups in North America, Australia, and New Zealand. Barker and Beckford (2006) used the term *indigenous technical knowledge* (ITK), which they used interchangeably with indigenous knowledge (IK). Beckford and Barker (2007) used the term *local knowledge* mainly to avoid the use of terms like *traditional* and *indigenous* in light of questions about the extent to which practices can claim to be untainted and purely traditional or indigenous. However, we would consider the terms most popular in contemporary use—indigenous knowledge, traditional knowledge, and local knowledge—to be interchangeable for this discussion.

Research dating back to the 1980s established the value and validity of traditional or local knowledge and skills in small-scale farming systems (Chambers, 1983; Chambers et al., 1989; Richards, 1985; Hills and Iton, 1983). Robert Chambers's work was seminal in the evolution of a new rural development paradigm founded on the general principle of putting farmers first (Chambers, 1983, 1997). Chambers concluded that the Green Revolution philosophy of knowledge transfer from the industrial to the developing world was inappropriate for resource-poor farmers in the global South. Richards (1985) showed how experimentation, innovation, and adaptation were constants among small-scale farmers—a process he called indigenous experimentation. According to Barker and Beckford (2006), indigenous knowledge and skills underpin these problem-solving efforts. Other recent research has confirmed the significance of local knowledge in smallholder farming (Agrawal, 1995; Beckford and Barker, 2007; Carswell and Jones, 2004; Pretty et al., 1999). Beckford and Barker (2007) make the point that the success of small-scale food farming is dependent upon farmers' traditional or local knowledge at the farm and village levels. They emphasize that such knowledge underpins the choices farmers make in farming techniques and provides strategies for soil fertility management, integrated pest management, crop selections, and a host of other micro- and macro-decisions.

The research of Richards (1985) and Chambers (1983, 1997) was instrumental in a reappraisal of the validity of local knowledge systems in tropical smallholder farming. This opened up new avenues of applied research in farming systems in Africa, Asia, Central America, the Caribbean, and South America, highlighting the role of traditional knowledge. Farmers constantly experiment, resulting in the development of homemade solutions to problems (Chambers, 1989, 1990; Chambers et al., 1989; Scoones and Thompson, 1994). Many traditional farming systems reflect complexity influenced by generations of observation and experimentation (Eyzaguirre, 2001; Shepherd, 2001) and producing food from diverse agro-ecological environments. It has become clear that these food production systems are far more sophisticated than might initially seem to be the case, and they contribute significantly to food security from the household level and upward.

The credibility of local knowledge and traditional farming skills in developing countries received a boost after the Rio Summit, which called for greater attention to be paid to the situated and contextual knowledge and experiences of local peoples in the developing world. In the past two decades or so, there have been significant institutional developments in the dissemination of information about indigenous knowledge through institutions like the Netherlands Organization for International Co-operation in Higher Education/Indigenous Knowledge (NUFFIC/IK-Unit) and the World Bank, which developed a program called Indigenous Knowledge for Development.

According to Briggs (2005), the research on traditional knowledge often focuses on its empirical and practical aspects, which in his view undermines a more comprehensive understanding of the epistemology of indigenous knowledge. He concedes the importance of the empirical, but argues that the emphasis on the empirical disconnects local knowledge from its context and ignores the economic and sociocultural contexts in which such knowledge is used. Kapoor (2002) supports Briggs's assessment, stating that the empiricist focus hinders a more rigorous theorization and politicization of local knowledge (cited in Beckford and Barker, 2007). Questions about the usefulness of local knowledge outside the spatial locales from which they originate (Briggs, 2005; Briggs and Sharpe, 2004), and its advocacy as a replacement for Western science, have also been criticized (Cleaver, 1999). In light of these critiques, it is important to situate local knowledge in the sociocultural, economic, political, and environmental contexts in which it developed and applied in Caribbean food production systems, thereby providing a framework for analyzing and assessing its role and value.

It is important to emphasize that local knowledge is not presented here as a panacea. Neither do we make any case for its superiority over Western scientific knowledge. Indeed, the dangers of romanticizing or idealizing traditional knowledge are well made in the literature (Beckford and Barker, 2007; Briggs, 2005; Forsyth, 1996; Sillitoe, 1998). Rather we make the point that much can be gained by seeing traditional knowledge and Western academic science as complementary ways of knowing, and acknowledge that local knowledge despite its limitations has evolved as a sound adaptation to local realities and is perhaps the single most important factor in the survival of the small-scale food production sector in the Caribbean and other parts of the tropics. We argue that local knowledge in the Caribbean is a necessary adaptation to well-recognized and well-documented resource constraints and institutional biases against the small-scale food production sector. The relevance of local knowledge must be assessed in terms of the structure of Caribbean economies and the agricultural sectors more specifically and the resultant institutional biases against the small-scale farming sector. It must also be assessed against the background of risks and uncertainties in challenging agro-ecological environments.

13.3 Local Knowledge and Rural Livelihoods: Examples of Local Knowledge in Caribbean Small-Scale Food Production Systems

There is much to be gained from investigating and documenting local knowledge that is of relevance to development problems (Briggs, 2005). Beckford and Barker (2007) investigated local farmers' knowledge in Jamaica. Through an analysis of empirical and conceptual issues relating to small-scale yam producers in central Jamaica, they highlighted the value of local knowledge in negotiating risky and uncertain environments and illustrated how small farmers are adaptive and innovative in their survival strategies and rural livelihoods.

13.3.1 Mixed Cropping and Symbiotic Crop Combinations

Caribbean small-scale farmers have demonstrated a sound knowledge of beneficial crop combinations and have applied this knowledge with great success for generations. Their informal knowledge of the symbiotic relationship between certain crops is derived from generations of careful observations and farm-level experimentations. Farmers will tell you that certain crops "go together" or "agree with each other" or that certain crops "are friends." This means that they provide reciprocal benefits and do well when planted together in mixed cropping systems, which characterize smallholder farming in the region. Other crops "do not agree" and should not be planted together. Their knowledge of complementarity among crops is quite sophisticated and takes into consideration ecological factors such as shade and exposure, control of weeds and pests, maintenance of soil fertility, variations in soil fertility, and spatial characteristics. In the Rio Grande Valley, Jamaica for example, dasheen is planted between bananas and plantains to make maximum use of fertilizers and receive shade from these two taller plants (Davis-Morrison, 1998). Some pests affect plantains, but not dasheen. Farmers efficiently use labor by weeding plantains and bananas when they are harvesting dasheen. Peas and beans go well with most crops and are planted in St. Vincent and the Grenadines on land that requires replenishing as peas and beans are known to have great nitrogen-fixing properties. Farmers throughout the Caribbean incorporate legumes into their cropping systems as a deliberate attempt to make use of their high nitrogen content. Throughout the region, peas and beans are grown with a variety of different crops, especially corn, which grows taller and provides shade. In Jamaica, gungo peas (pigeon peas) are grown with cassava and pumpkin. The cassava and pigeon peas provide shade, while the pumpkin provides ground cover. In the yam-growing regions of Jamaica, yams are planted with coco (coco yams) in the main yam-growing regions of the country. In other farming systems like the Rio Grande Valley, yellow yams are not planted with dasheen (a crop similar to coco yam) because farmers say they require the same nutrients and are therefore in competition (Davis-Morrison, 1998). Cabbage is not planted with coco yam and dasheen as they shade the cabbage and prevent it from folding. Peppers and the sorrel do not go together as they impede each other's growth (Davis-Morrison, 1998). Mixed cropping is also a method of agricultural intensification and an adaptation to limited land by small-scale farmers.

Farmers also use their ethnobotanical knowledge of plants to identify problems. For example, in Grenada, certain plants are used as indicators of diminishing soil fertility and the need to rest land. In Dominica and Grenada, the occurrence of certain plants is an indication of the presence of certain pests. Farmers in south St. Elizabeth, Jamaica, use certain plants to gauge soil moisture availability. Other plants indicate soil and environmental properties that could represent beneficial micro-conditions for certain plants.

Knowledge of good farming practices is constantly being accumulated over time. There are examples of the use of traditional cultivation techniques like mixed cropping by small-scale farmers from all periods of the Caribbean's settlement history: the pre-Columbian period (Sauer, 1966), food production by plantation slaves (Mintz, 1985), and in the post-emancipation period (Satchell, 1990). The classical agronomic research on crop combinations in the Caribbean was done in Christiana, Jamaica, between the 1950s and 1980s (Innis, 1997). Further evidence of successful polyculture in the region is found in the studies of the highly productive, agrodiverse Caribbean kitchen gardens or food forests (See for example, Brierley, 1976, 1991; Hills, 1988; Hills and Iton, 1983; Thomasson, 1994).

13.3.2 Traditional Knowledge and Soil Management

Caribbean farmers rely heavily on their traditional knowledge of soils in specific locales to make decisions about their cropping systems. Examples throughout the region demonstrate that local farmers are knowledgeable about variations in soil type and the physical properties of soils in localized areas. For example, in the Millbank area of the Rio Grande River Valley in Jamaica, farmers differentiated and classified soils on the basis of crop yield, soil texture, and soil color (Davis-Morrison and Barker, 1997). Beckford and Barker (2007) suggest that crop yield is a ubiquitous, somewhat tautological, category used by farmers to identify soil type. Farmers also use the smell of the soil and the presence of micro-organisms such as earthworms to gauge fertility and acidity. Soil texture is described by various terms such as "soft," "stiff," "sandy," "gravelly," "gummy," and "tough." In St. Vincent, similar terms were used for heavy soils and soils with high clay content, described as sticky and unsuitable for certain crops, especially onions (Collymore, 1983). Farmers test soil by rubbing it between their fingers, by smelling it, by inserting a machete and some older farmers have been known to taste soil as well. Signs of land degradation are also well known to farmers and are used to determine when to rest land or rotate crops (Bailey, 2003).

13.3.3 Traditional Knowledge in Decision-Making

Decision-making among small-scale farmers in the Caribbean and throughout the tropics is greatly influenced by traditional knowledge. Smallholder farmers assess the advantages of particular courses of action against the background of their knowledge of local environments and economy. For example, the failure of agricultural authorities in Jamaica to achieve more than negligible adoption of minisett technology by commercial yam farmers is largely due to farmers' assessment that the system is not well suited to local soil, topographical, and market conditions. This assessment is informed by their local knowledge informed by their lived experience. As an example, yam farmers in the prolific yam growing Trelawny Parish argued that the use of plastic mulch to cover continuous mounds planted with yams would not work in the area because of the heavy clay soils. Farmers opined that in the tropical conditions and under the cover of this plastic mulch, clay soils would become water logged, leading to rotting of yam plants (see Beckford, 2000, 2002). Farmers' knowledge of topography and slopes also led them to conclude that minisett would not do well on certain slopes without the use of yam sticks—which the minisett system does not employ.

Research elsewhere in the Caribbean emphasizes the role of traditional knowledge in decision-making among small-scale farmers. For example, in 1983, farmers in the north of St. Vincent were invited to expand or start onion production. Many farmers who did not get involved cited the unsuitability of the soils in the area for onion cultivation as their main reason. In their assessment the soils on their farms were too "sticky"—a reference to the high clay content in soils in the area—and poor soil drainage would cause the crop to rot (Collymore, 1983).

13.3.4 Traditional Knowledge and Sustainable Agriculture

Local knowledge also plays a big role in sustainable agriculture in the Caribbean. All the examples we discussed in the preceding chapter are significantly influenced by traditional knowledge and skills. The farm- and village-level experimentations of small-scale farmers around the region can be seen as homemade on-farm adaptive research grounded in traditional knowledge as well as extending it and creating new knowledge. Many elements of sustainable practices have formal academic credibility, but among the small-scale farming sector, they developed over time and are intergenerational. Examples include the practice of mixed cropping and crop rotation, which have been practiced in the region since at least the slave era.

The use of grass mulch as an adaptation to moisture deficiency, recycling of yam sticks, farm fragmentation, use of organic manures, and integrated pest management are all grounded in the individual and collective knowledge that small-scale farmers developed overtime in informal settings and that passed on orally and through practice and observation from generation to generation. These adaptations reflect the ingenuity and resourcefulness of farmers in the region.

13.3.5 Traditional Knowledge and Natural Hazards

The unequivocal nature of climate change and the desire of vulnerable groups to cope with these changes have led the IPCC (2007) to place local knowledge at the epicenter of discussions on adaptation options. Farming communities have long used local knowledge and fine-tuned their farming practices to local environmental conditions, including weather and climate variations. Jones et al. (1999) argued that planned adaptation to future climate will be based on current individual, community, and institutional behavior developed in part as a response to present climate, an argument supported here. Farmers' perceptions of climate are inseparable from the rich body of local knowledge, which has helped them to cognize and negotiate numerous environmental uncertainties. Farmers may not understand the science behind climate change, but because observing the environment is crucial to their livelihood, they are well in tune with the weather and as such are well positioned to discern even minor deviations from what are considered normal conditions.

The congruence of farmers' perceptions of climate and meteorological record has been noted elsewhere (Vedwan, 2006; Gbetibouo, 2009) amidst growing interest in bridging the gap between traditional and scientific knowledge in light of the unequivocal nature of climate change (IPCC, 2007) and the desire to enable vulnerable groups to adapt. The analysis of the intersection between local and scientific knowledge offers a unique opportunity to further understand and synergize the relationships between the two areas. Gyampoh et al. (2008, p. 71) proposed that, as far as climate change is concerned, local and scientific knowledge should be complementary on the basis that

> models and records of precipitation mainly focus on changing amounts of precipitation with climate change, [while] knowledge of indigenous people also emphasis changes in the regularity, length, and timing of precipitation.

In other words, local peoples take a more nuanced look at climate change and draw conclusions and fashion responses, which are fundamentally different from scientists' approach. Studies indicate that, although farmers may not understand the science behind climate change, their continual observation of and interaction with the surrounding environment heightens their awareness of even minor deviations from what is perceived to be normal weather and climate conditions.

13.4 A Rational View of Traditional Knowledge

Research has highlighted the role and value of local knowledge in the context of small-scale food farming in the Caribbean. Our examples and discussion should not be interpreted as an unqualified endorsement of traditional knowledge. Briggs (2005) reminded us that traditional knowledge is not without its problems. A major criticism of local knowledge is that it tends to be location specific and its relevance and usefulness outside its place of origin are therefore questionable (Kalland, 2000). Briggs (2005) also questioned the transferability and universal value of indigenous knowledge. Beckford and Barker (2007) posit that if the concern is about transferring specific examples of local knowledge to contexts in which they were not developed, then this is a valid question. They argued, however, that the aim should not be to transplant specific traditional knowledge and practices to other areas, but to transfer the principle(s) of traditional knowledge and try to find local examples that would be relevant, appropriate, and useful in specific contexts. Nevertheless, Briggs's caution is instructive. It is the context-specific relevance of traditional knowledge and praxis to an area that makes it useful. Decontextualize it and it is no good. Interestingly enough, the same argument could be made about Western scientific knowledge and its products, many of which have been miserable failures in places where they have been transplanted but to which they were unsuitable hence inappropriate. Some researchers have suggested that if isolated from its cultural context and forced into a Western scientific academic epistemology, traditional knowledge could be emasculated in the eyes of local people (Jama, 1987; Thrupp, 1987).

The dangers of romanticizing traditional knowledge have been discussed by a host of writers including Beckford et al. (2007), Briggs (2005), Forsyth (1996), and Sillitoe (1998, 2010). Briggs argues that it is problematic to perceive and present traditional knowledge as an untainted, pristine knowledge system. Local knowledge will not always provide sustainable solutions to rural problems (Briggs, 2005). Our own research bears out this view. As Barker (2010) points out, Caribbean small-scale farmers improvise and experiment daily on the basis of their local knowledge but not always successfully. Briggs (2005) also warned against presenting traditional knowledge as a unitary knowledge shared more or less equally across a community. This is a good observation, as in our own research we have seen how perceptions, beliefs, and practices can vary greatly over small spatial contexts.

Throughout the tropics, traditional knowledge and praxis are mainstay of small-scale farming systems. But domestic food production continues to struggle and food security in many parts of the tropics is actually worsening. It is therefore legitimate to ask as a Tanzanian farmer did (see Briggs, 2005): why is small-scale farming so unsuccessful if traditional knowledge is so good? According to Beckford et al. (2007), the answer might simply be that there are numerous other factors at play, over which small-scale farmers in general have no control. However, the question does emphasize the danger of presenting traditional knowledge as a panacea and underscores the importance of tempering our expectations of what local knowledge can do and presenting a balanced account of its application in Caribbean small-scale farming systems. In the interest of balance, it has to be conceded that these applications are not always unproblematic. Take, for example, the perceptions of the risk of agrochemical use among small-scale farmers in St. Vincent (Grossman, 1992, 1998), Grenada, and Jamaica (Semple et al., 2005). Farmers clearly put themselves and perhaps others at risk with their use, storage, and disposal of agrochemicals and their incomplete and inaccurate knowledge of proper procedures. We have also seen the cultivation of steep slopes leading to land degradation, especially soil erosion, and while it is inaccurate and unfair to suggest that soil erosion and land degradation are endemic to only the small-scale farming sector, it has to be said that farmers sometimes engage in practices that are unsustainable.

At a more philosophical level, the question should be asked: to what extent is any farm practice or piece of knowledge truly or purely indigenous, local, or traditional? In other words, is it possible to disentangle the influence of science in some examples of local knowledge? Take the example of our innovative minisett yam farmer whom we discussed in Chapter 12, who successfully modified the minisett yam package. He was exposed to the system through his part-time employment with the Forest Department and his system is clearly influenced by Western scientific knowledge. As Beckford et al. (2007) explain, small-scale farmers do have contact with the scientific community through a variety of media, and so it should be expected that there would be some modern scientific influences in some of their practices. We would suggest that what we typically describe as traditional is in many cases more hybridized knowledge sources than is at first recognized.

Some writers have suggested that traditional knowledge and Western science are incompatible (Mohan and Stokke, 2000), but there now seems to be a critical mass of academic thought that concludes that there can be intersections between them (Emeagwali, 2003; Lado, 2004; Liwenga, 2008; Marin, 2010; Mercer, 2009; Mercer et al., 2010; Reed et al., 2007). Our research suggests both the bipolar tensions between the two systems as suggested by Briggs (2005) and the potential and actual synergies between them.

13.5 Traditional Knowledge as a Survival Algorithm in Caribbean Small-Scale Farming Systems

The ability of small-scale farmers in the Caribbean to successfully navigate challenging, sometimes harsh, agro-ecological environments is derived from their knowledge of these environments, which has allowed them to devise adaptive and coping responses through observation and experimentation. The example of farmers on the southern coastal plains of St. Elizabeth Parish in Jamaica is a perfect manifestation of this point. The plains of south St. Elizabeth are hot, with an uncompromising dry climate characterized by annual precipitation ranging from 650 to 800mm of rain, which in some parts of the world would be described as a semi-arid condition (Barker, 2010). However, this part of St. Elizabeth is known in some quarters as the bread basket of Jamaica as it is one of the principal areas for the production of domestic food crops, which can be found in the fresh produce markets that characterize large, small, and medium-sized towns across the country and in resort facilities in Jamaica's world famous tourist resort towns—Negril, Montego Bay, and Ocho Rios. While the area grows an impressive range of staple domestic food crops, it is especially famous for fresh fruits and vegetables such as tomatoes, carrots, lettuce, cabbage, zucchini, broccoli, cauliflower, a variety of melons and cantaloupes, a variety of peas and beans, and condiments such as skellion, thyme, parsley, and peppers.

The farming environment is fragile and hazard prone, with drought and hurricane as unwelcomed, but integral, parts of the agro-ecological milieu. But farmer ingenuity has enabled them to survive and even thrive and achieve impressive levels of production and develop resilience in the farming system in an environment that is marginal at best. Farmers in the tropics utilize mulching as an adaptation to low rainfall. Without it, farming in south St. Elizabeth would be unsustainable, and it is thus an integral part of farmers' arsenal of traditional knowledge (Barker, 2010). Mulching suppresses weeds, thus reducing labor input and the need for chemical weed control. Grass mulch residues are eventually incorporated back into the soil to improve organic content and soil fertility. Farmers also say that mulch acts as a carpet for crops like melons and cucumber to rest on thereby reducing scars and blemishes due to *soilburn*. Farmers combine mulching techniques with improvised irrigation to help cope with low-rainfall conditions. It should be noted that grass mulching is used in farming systems in other semi-arid parts of the tropics. In the village of Kititimo, in Singida Region in Tanzania, villagers use grass mulch for small home garden vegetable plots to retain moisture. Grass mulch is used extensively in the vegetable nursery beds. After seeds are sown, a layer of grass mulch is put down on the raised seed beds before water is applied. This is an example of the application of the traditional knowledge about micro-climates, soils, and crop needs in two different parts of the world with different cultures, which puts into perspective one of the main criticisms of indigenous knowledge, namely that it is useless outside its geographical origin. Indigenous knowledge and practice are, in fact, transferable where there are locations with similar conditions and needs.

Although Caribbean small-scale farmers draw heavily on indigenous knowledge in their farming practices, their skills and knowledge are seldom discussed within the conceptual framework of indigenous knowledge. Regardless, traditional knowledge is a normal part of daily life among small farming communities in the Caribbean as in other parts of the tropics (Barker and Beckford, 2006). Indigenous knowledge is a repository of community-based empirically derived data on the environment, ecology, culture, and social traditions, and it provides a flexible framework for problem-solving (Barker and Beckford, 2006, p. 537).

In this chapter we argue that honoring and supporting local or traditional agricultural knowledge and practices should be part of any strategy to enhance food security in the Caribbean. Local knowledge here is defined as a complex body of know-how, practices, and skills that are developed and sustained by communities with shared experiences and histories. In the Caribbean such knowledge provides a framework for decision-making on social, political, economic, and environmental issues. In the area of agriculture, such knowledge has been shaped and modified by ongoing farm-level experimentation over many generations (Beckford and Barker, 2007). It has developed outside the formal education system, is embedded in culture, and is steeped in tradition and is the basis for decision-making in rural communities with respect to food security and natural resources management (Beckford and Barker, 2007). Local knowledge is localized in nature and traditional in context. It is often unique to specific agro-ecological and cultural contexts, but adaptability to other contexts is also often possible. It is constructed informally, transmitted orally, and is hardly ever documented by farmers.

The validity and value of traditional skills and knowledge in small-scale agricultural systems have long been recognized (see Chambers, 1983; Chambers et al., 1989; Richards, 1985; Hills and Iton, 1983). In the Caribbean, local knowledge has evolved as a sound adaptation to local realities. It is a necessary adaptation to resource constraints in a challenging agro-ecological, cultural, economic, and political milieu rife with risks and uncertainties (Beckford and Barker, 2007). Successive generations of small-scale food farmers in the region have successfully used traditional agronomic practices related to soil management and conservation, disease and pest control, moisture retention, crop combinations, and cropping systems, among other things. Small-scale food farmers in the Caribbean have traditionally operated in an environment where they are fending for themselves and depending on their own ingenuity to make a livelihood. In this context, local and traditional knowledge and practices are indispensable to the survival of the domestic food production sector, rural development, and food security. Indigenous knowledge is not a static body of 'traditional' knowledge but the creation of knowledge through a dynamic process with content that is responsive to change from a variety of sources (Barker and Beckford, 2006).

Small-scale agriculture and traditional knowledge have received harsh criticism about their value and ability. Of note are the views of Collier (2008, 2009), who has argued that small-scale agriculture is ill-suited to the demands of modern agriculture in which scale is important and which he argues must be driven by scientific agriculture fully integrated into a market system (Collier, 2009). This view has been challenged (see Wiggins, 2009), but emphasizes that indigenous knowledge still remains marginalized in practice.

The way traditional knowledge is positioned might be at the heart of some of the negative critiques. It is sometimes presented as content, as acquisition of knowledge, or as a process of learning. This last view has received some advocacy in the recent literature (see Berkes, 2009b) and makes it important to keep the local and cultural contexts in mind when local knowledge is being analyzed (Berkes, 2009; Dixon, 2005).

13.6 Summary

We argue here that local knowledge in Caribbean small-scale farming systems has to be assessed as a framework for survival and livelihood security in an agro-ecological environment rife with risks and uncertainties. The dual structure of the agricultural sector in the region has led to a host of institutional biases against the small-scale farming sector. These are manifested in inequities in terms of access to credit and loans, poor physical infrastructure, lack of research into domestic crops, and poor infrastructure for marketing and distribution of produce. In addition, hazard relief for farmers is slow, disorganized, and inadequate and does not build adaptive capacity and resilience. The liberalization of trade has exposed local farmers across the region to virtually unfettered unfair competition from cheap food imports produced in heavily subsidized production systems. Pretty et al. (1999) posited that small-scale farmers in the tropics are typically remote from agricultural services. With very little state support, small-scale farming survives because of the ingenuity of farmers who have constructed knowledge from different sources, which draws on as well as cater to their lived experiences. In this context, traditional knowledge is more than just valuable. It is an invaluable and indispensable part of rural livelihood and household and national food security. In the Caribbean context, farmers' knowledge and skills underpin survival strategies (Barker, 2010).

Farmers' reliance on local knowledge is critical to the survival of small-scale farming systems in the Caribbean and is key to successful food production and hence food security. We have discussed some of the persistent and seemingly intractable structural problems associated with dualism and dichotomized structure of Caribbean economies, disasters and globalization, and their impact on food production and food security. In this context, farmers' local knowledge is a *sine qua non* for successful local agriculture.

PART IV

Globalization and Caribbean Food Security:
Challenges and Opportunities

CHAPTER 14

Food Imports, Domestic Production, and Food Security in the Caribbean

14.1 Overview

Food security is determined by local production, agroprocessing, food aid, food trade, and food reserves or stockpiles. We have already seen that local production in the Caribbean has declined significantly over the past two to three decades, and CARICOM countries as a whole have become net importers instead of net exporters of food. The region is now very heavily dependent on food imports to meet its food needs (Beckford and Bailey, 2009; Deep Ford and Rawlins, 2007; CFNI, 2007; Timms, 2008; Potter et al., 2004; Weis, 2004). In 2006, for example, Jamaica imported some US$1.64 billion worth of food, which was half the country's total import bill (Beckford and Bailey, 2009). The situation is similar in many other CARICOM countries and the developing world where markets have been opened up through trade liberalization (Short, 2000; Spitz, 2002; Walelign, 2002).

We would argue that this dependence on imported food constitutes a major threat to Caribbean food security. First of all, purely from a livelihood perspective, it does immeasurable damage to local producers and rural development. Faced with unfair competition and the dumping of cheap, heavily subsidized food mainly from the United States, many farm families experience difficulty obtaining a satisfactory livelihood for themselves (Beckford and Bailey, 2009; CIOEC, 2003; Via Campasina, 1996, 2001, 2003; UNDP, 2005). Most of the imported food to the Caribbean comes from the United States, where heavily subsidized production enables that country's farmers to sell for less than the cost of production (Windfuhr, 2002, 2003; Windfuhr and Jonsen, 2005). Caribbean farmers are therefore forced into unfavorable, often insurmountable, competitive situations and in Jamaica, for example, many have succumbed to this dumping of cheap exports and have gone out of business (Beckford and Bailey, 2009).

Conventional wisdom would suggest that *ceteris paribus,* opening up local markets to international competition, should result in lower prices and thus is beneficial to consumers and that this competition should stimulate more efficient local production, thereby providing even greater access to affordable food. The problem is that all things are not equal. Unfettered competition from heavily subsidized foreign food producers has coincided with the removal of subsidies from local producers, creating an uneven playing field. The high dependence on imported food raises other obvious dangers as well. For one thing, the structure of the world economy means that external shocks often reverberate throughout the system with devastating consequences for the most vulnerable nations and people. Market stability is a major concern here, which causes uncertainty in supplies and increasing prices, which could both result in food shortages in the region. In light of the dependence of the Caribbean on food from the United States, terrorist attacks on the food system in America could have serious implications for Caribbean food security. The US$26 million Food Security Project implemented in 2002 was partly in response to the near-food crisis in some Caribbean countries in the aftermath of the September 11 terrorist attacks on the United States. Food safety is an important part of food security, which perhaps does not get enough attention in the literature on food security. In the context of the Caribbean, we would suggest that this should be a real concern. Apart from the threat posed by terrorist attacks, the long time taken for imported food to reach from one place to another increases the risk of contamination of food (Halweil, 2005). Beckford and Bailey (2009) pointed out that the longer the food travels, the more it changes hands, which increases its risks for contamination. Halweil (2005) argues that the centralized nature of American food production and processing increases the risks of contamination of food, while the sheer size and uniformity of farm operations create ideal conditions for the rapid spread of diseases. The dependence on foreign food and especially American food is therefore a cause for concern.

This chapter analyzes the issue of regional dependence on imported food and makes a case for strategies geared toward greater food self-sufficiency in the region.

14.2 Food Security and Domestic Food Production

Domestic food production and imports are the principal components of food availability in a country. While there is a paucity of studies specifically on the problem of declining food production there are numerous studies that describe the socioeconomic trends affecting agriculture and food output in the region (Semple, 1996 cited by Campbell, 2011). The performance of the domestic food production sector on a whole has been found elsewhere to be linked to patterns of food insecurity and vulnerability (Shapouri and Rosen, 1999; Ghatak and Ingersent, 2002).

Ghatak and Ingersent (2002, p. 5) posit that

> in a country with a lagging agricultural sector and a large unmanaged food import bill, it might well make better economic sense to expand food production for the domestic market than to encourage agricultural exports.

The domestic food production sector contributes approximately 90 percent of the food consumed in most food-insecure countries (Shapouri and Rosen, 1999), yet the importation of food is increasingly becoming the preferred option for food availability. Conway (1998) and Beckford (1972) argued that domestic food production sectors are poorly integrated into the local markets and traditional practices have favored imported foodstuff; hence, the domestic provision of food has rarely been a major objective of Caribbean agriculture.

Previous research on domestic food production in the Caribbean has focused mainly on the factors that have contributed to its decline over the years. Chief among these is the work of Semple (1996 cited by Campbell, 2011), which observed that the literature is fraught with perspectives from political economy and political ecology. At different levels, these authors assessed how power and class relations have the potential to "distort the levels and direction of agricultural development in these countries" (Watts, 1983, p. 7).

The political economy and ecology theses came under heavy challenge from Semple (1996 cited by Campbell, 2011), who found their usefulness for policy prescriptions to be limited. According to Semple, most writers (e.g., Watts, 1983) called for radical social and economic transformations that the majority of developing countries simply do not have the political will to implement. Semple therefore recommends that in order to conceptualize declining food production in the Caribbean and simultaneously avoid sterility in policy prescription, a structural approach should be utilized with explanations of the complex ways by which human beings interact with their physical environment sought in the nature of the physical environment, as well as the dominant socioeconomic forces shaping society.

A general consensus in the domestic food production literature is that economic structures and relations set up during colonialism have persisted in the post-independence period, so that most of the region's agricultural resources are still being used to support export agriculture (Semple, 1996 cited by Campbell, 2011). Local food production continues to be compromised because small farmers are still obliged to occupy marginal lands, find difficulty in obtaining farm credit, or lack proper infrastructure, such as feeder roads, storage facilities, farm subsidies, and extension services (FAO, 1988).

Social processes such as urbanization have also shaped domestic food production in the region. According to Semple (1996, p. 6), "increased urbanization in the Caribbean has led to changing consumer habits, such as preference for 'ready to cook' and 'ready to eat' foods." As a result, there was extensive importation of such convenient foods throughout the region. Ahmed and Afroz (1996) blame this preference on the penetration of foreign culture and commercial advertisements by the local and overseas media. They posit that over the years the population has been conditioned to accept imported food.

In terms of agricultural land-use changes, an increase in the amount of farmland left idle is one reason for declining agricultural output in the region (Brierley, 1988). The increase in idle land has been related to the widespread perception among youth that agriculture represents a low-income activity (Semple, 1996 cited by Campbell, 2011). In Jamaica, further degradation of marginal agricultural lands used for domestic food production is a critical land-use-related problem affecting food output (McGregor and Barker, 1991).

The FAO (1999, p. 2) defines food security as the condition that exists when "all people, at all times, have physical and economic access to enough safe and nutritious food to meet their dietary needs and food preferences for an active and healthy lifestyle." Central to the concept of food security, therefore, are issues relating to food availability, affordability, and utilization (Jenkins and Scanlan, 2001). Further, the state of poverty, disasters, trade, and the general condition of agriculture in a country are also important food security issues (Sen, 1981; Trueblood and Shapouri, 2001; Shipton, 1990; Pelling and Uitto, 2001). While domestic food production remains a central component of food security in developing countries, trade liberalization and structural adjustment policies have contributed to increased vulnerability to external shocks in the global economy.

Food availability at the national level can be increased either by accelerating domestic production or by increasing food imports. In areas where domestic agricultural productivity is limited, food imports should play a major role in the country's food security (Trueblood and Shapouri, 2001). However, the authors identified a gradual increase in the dependence on imported food in many developing countries that have the capacity to increase domestic production. They argue for more recognition of the important role of domestic production in food security.

14.3 Global Change and Agriculture in the Caribbean

Generally, the Caribbean is a net importer of food, "a paradise that cannot feed itself" (Ahmed and Afroz, 1996, p. 4). There is no real consensus as to when the decline in agriculture may have started in the Caribbean; however, Ahmed and Afroz (1996) suggest that the turning point for Jamaica may have been after independence and the problem may have intensified during the 1980s under structural adjustment policies (SAPs). In 1981, according to Ahmed and Afroz, there was a 40 percent increase in food imports, due to the removal of import restrictions on a range of foodstuffs that could be produced locally. This resulted in a 12 percent reduction in domestic agricultural production the following year, because of competition from cheaper imports. Weis (2004, p. 463) described this decline in domestic food production in Jamaica as a crisis, arguing that

> Jamaica's agricultural crisis is reflected in a soaring agro-trade deficit; from a near balance in the early 1990's, agro-exports now equal only 60 percent of agro-imports. Jamaica's agro-imports erupted in the 1990's following liberalization. Between 1991 and 2001 Jamaica's total food and beverage imports increased by two-and-a half times, from US$ 199 million to US$ 503 million.

This argument is also supported by Ahmed and Afroz (1996, p. 7), who submit that "[n]ot only has Jamaica been unable to expand its agricultural items, but it faces a major food crisis because of declining production of staple food for the local market."

However, despite the decline, Weis (2004, p. 485) believes because agriculture provides more direct employment than the other sectors in the island, it is a "vital sponge for labor" and an important tool for development. This argument is supported by Hawkins (1979), who argued that agriculture is the lifeline of many Caribbean countries, yet is largely neglected by Caribbean decision-makers. Mellor (2002, p. 1) effectively sums up the issue: "a key theme that emerges is that agriculture potentially benefits more proportionally than other sectors but also suffers more from constraints to benefiting."

Economic opinion suggests that the decline has become a feature of Caribbean agriculture in recent decades. Weis (2004, p. 463) challenges this notion by postulating that the decline of Jamaican agriculture is not inevitable, but requires a radical imagination supported by political will. In the past decade or so, this decline in agriculture has occupied center stage in discussions relating to domestic food production. Caribbean small-scale farmers continue to face common challenges brought on by a changing economic and political environment. The current state of domestic food production and food security in the Caribbean is a product of the combination of global economic and environmental forces, as well as local institutional changes. Environmental and economic changes occur in concert to shape vulnerability, domestic food production, and food security, and local institutional changes and processes can shape coping and adaptation strategies, which in turn affect domestic food production and food security. Local knowledge is also seen to have a direct impact on coping and adaptation strategies, which affects rural livelihoods in general (Beckford, 2007; Campbell, 2011).

Over the past two to three decades, CARICOM countries as a whole have been experiencing declining domestic food production. The region is now very heavily dependent on food imports to meet its food needs (Beckford and Bailey, 2009). In 2008 the Caribbean's food import bill was estimated at US$ 3.5 billion (CARICOM Statistical Office, cited in Regional Food and Nutrition Security Policy, 2010). Caribbean countries are not able to produce all their food, but most countries are capable of producing most of their food requirement, thus reducing their food import bill, decreasing their dependence on imported food, and reducing their vulnerability to external shocks.

14.4 Globalization and Agriculture

Defined as the expansion of economic, social, and political institutional forces across borders, globalization has left its mark on almost every aspect of Caribbean life, especially within the agricultural sector. The process started in the Caribbean in the early 1980s (Ahmed, 2001) and manifested itself in two ways. Initially, food crop subsidies for local farmers were gradually terminated, leading to higher production cost and consequently higher prices for domestic produce. This was followed by the liberalization of local markets (Ahmed, 2001), resulting in significant increases in imported food.

The liberalization of domestic markets was intended to increase competition, productivity, and the economic situation of farmers across the world (WTO, 1997). On the existing evidence, however, it is felt that the disadvantages of trade liberalization outweigh the benefits for developing countries. Shiva (1997, p. 5) observes that "in country after country the process is leading to a decline in food production and productivity, a decline in conditions for farmers and a decline in food security for consumers." Research in the

Caribbean has shown that small-scale farmers will be among the biggest losers from trade liberalization processes (ECLAC, 2004). Capacity-building initiatives grounded in the local experience and realities of these farmers are critical to the viability of the sector, rural livelihoods, and sustainable development. The main danger of globalization to the agriculture sector is in the pace of the process (Ohga, 1998). The complex and dynamic nature of trade liberalization processes has also resulted in varying levels of impact in the small farming sector. It is argued that small-scale farmers were unprepared for the process of globalization, which resulted in devastating impacts (Ahmed, 2001). This is not surprising given the unsophisticated nature of small-scale farmers and the institutional isolation and marginalization they face. Some countries have adjusted to trade liberalization processes through the transfer of agriculture technologies and the replication of sectoral "best practices" (Ohga, 1998). However, this approach could result in a displacement of vast numbers of workers from the agriculture sector (Ohga, 1998).

In the Caribbean, cheap food imports have resulted in increased competition in the domestic market and pose a significant threat to the livelihood of small farmers who often have limited access to critical resources such as land and credit, which has crippled their ability to compete effectively with heavily subsidized food from developed countries (FAO, 2003; Ahmed, 2001; Ohga, 1998; Khor, 2006). Furthermore, SAPs have limited the capacity of governments in the region to provide support for the domestic crop sector.

On the positive side, Mellor (2002) postulates that globalization provides an opportunity for developing countries to reposition agriculture as an engine for economic growth. He posits that globalization "increases the potential for agriculture to increase food security through enlarged multipliers to the massive, employment-intensive, non-tradable rural non-farm sector" (p. 4). Additionally, trade liberalization may provide unique opportunities for farmers to engage in export marketing and agroprocessing sectors, which could push them further up the value chain. However, the reality is that small farmers often lack the capacity to take advantage of these opportunities. Hence, trade liberalization of agriculture is a major threat to domestic food production and food security. That said, it should be noted that there is tremendous untapped potential for Caribbean food and agricultural commodities to capitalize on the benefits offered by globalization. There is a large and growing Caribbean diaspora community in North America and Europe yearning for authentic Caribbean food and other products. Like many other developing countries, the Caribbean has perhaps been more concerned with the negative consequences of globalization and has therefore squandered opportunities.

14.5 Imported Food and Caribbean Food Security

There is a close link between international trade and food security in the Caribbean. The following statement emphasizes the regional scenario:

> The region was seduced by the importation of cheap food and paid less attention to food security. . . . The soaring food and agriculture commodity prices in 2007/08 and financial crisis of the last few years caught the Region unprepared and put food security at the forefront of the regional agenda.

This quote is part of an assessment by the CARICOM representative to the World Food Summit in 2009. Imports constitute a major portion of agricultural trade in the Caribbean. Except in Guyana and Belize, agricultural imports constitute more than 60 percent of agricultural exports (Deep Ford and Khaira, 2007). In fact, the dependence on imported food in the Caribbean is higher than in developing countries as a whole (Deep Ford and Khaira, 2007).

The heavy dependence of CARICOM countries on imported food represents an issue of vulnerability, which is an important point of analysis in food security discourse. International trade now plays an increasingly important role in Caribbean food security. Deep Ford and Rawlins (2007) posit that trade and food security are linked together in the region. Emphasizing that most countries in the region are net importers of food, they argue that food imports are critical to food security, especially in terms of nutrition and stability of supplies. The Ministry of Agriculture of Jamaica emphasized the importance of food imports to the country by revealing that imports make up the difference between domestic production and consumption needs, reducing supply variability and fostering economic growth (MOA, 2001). But while food imports have indeed become fundamental to food security in Jamaica, as is the case in the rest of the region, the assessment by the Ministry of Agriculture gives the impression that increased food importation was stimulated by shortfalls in domestic production. This implication should be questioned. We would argue that the importation of some foods takes place at the expense of local producers and small-scale farming sector and disrupts the economic livelihoods of numerous rural farm families (Beckford and Bailey, 2009; McGregor et al., 2009; Deep Ford and Rawlins, 2007). Research shows that many of the foods imported by Jamaica have been successfully grown by successive generations of local farmers (Beckford and Bailey, 2009; Campbell, 2011). These studies have shown that faced with competition for cheaper imported food, many small-scale farmers have left farming or scaled back their production. Certain crops that were once grown successfully are hardly grown anymore (Beckford and Bailey, 2009).

Smallholder farmers in the Caribbean are forced to compete mainly with food from the United States. In Jamaica, trade liberalization led to a flood of imported food in the 1980s. Structural adjustment led to deregulation of local markets, removing protective barriers and exposing local farmers to virtually unfettered competition for imported food (Beckford and Bailey, 2009). This competition does not take place on a level playing field. Because more often than not this imported food is produced under subsidized conditions, farmers are able to sell at prices lower than the cost of production. This "cheap" food is therefore traded to Caribbean countries such as Jamaica under conditions referred to as "dumping." Caribbean smallholder farmers for their part operate with a number of fatal constraints identified earlier in this volume and in the general literature on agriculture in the region, which limits their ability to compete successfully.

The largest area of imports is cereals, accounting for almost one quarter of total food imports. The main cereals imported are rice, wheat and flour, and corn. In the past 12 years, cereals have accounted for an average of 80 percent of food imports in Jamaica. As a whole the region has the capacity to be self-sufficient in rice cultivation, but only Guyana and Trinidad and Tobago grow rice in significant quantities. Intraregional trade does not satisfy regional needs. This is because rice is the main staple food for much of the populations of these two countries and much of the rice output is consumed locally. Rice has been successfully experimented with in Jamaica but has not caught on among small-scale food-farmers even in ecologically favorable areas. Throughout the region, rice is a main staple food and plays an important role in Caribbean cuisine. Wheat is also a fundamental part of the diet of Caribbean peoples, but the region does not have the right combination of climate and topography for the successful cultivation of wheat on a large scale. The importance of wheat in Caribbean diets and cuisine can be traced back to the slave plantation era, when wheat flour was one of the key components of slave diets supplied by slaveholders. This continued after emancipation, and wheat flour is an important part of the world-famous Caribbean cuisine. There are very few meals prepared in the region that do not include wheat flour. In the 1970s the government of Jamaica promoted the idea of greater food self-sufficiency. There was a campaign to get Jamaicans to eat more locally grown food. Instead of wheat flour, cassava flour from locally grown roots was advocated. This effort met with very little success.

Corn is also an important food in the region. It can also be grown well throughout the region but is not cultivated in significant quantities except in

Belize. Elsewhere it is grown mainly as a rotational crop on very small plots. One reason is that successful commercial corn cultivation requires huge tracts of land and resources to cultivate it. Mechanization would also be required to ensure efficiency. Much of the local corn grown is used in the manufacturing of livestock and poultry feed. Locally grown unprocessed corn is also used in local cuisine while imported corn meal (grounded corn) serves more formal uses.

Importation of cereals, especially wheat, has little impact on local agriculture although the region could easily improve its self-sufficiency in rice and corn. However, for a host of other traditionally important cash crops, importation adversely impacts local agriculture. These crops include onions, Irish potatoes, carrots, kidney beans, lettuce, cabbage, and cauliflower. In the south St. Elizabeth area of Jamaica, many farmers have scaled back production or gone completely out of production of some crops due to this competition from imported foods. The Planning Institute of Jamaica reported that the production of red kidney beans in the island declined by 81 percent between 1994 and 2001 (Beckford and Bailey, 2009). Even higglers who traditionally marketed locally grown produce in public markets, which is common throughout the region, have now begun to buy and resell imported food like carrots, onions, and red kidney beans.

Increasing food insecurity and loss of rural livelihoods have been experienced by several Caribbean countries, and trade liberalization was a contributing factor as it resulted in the loss of access to markets for some traditional export crops (Deep Ford and Rawlins, 2007). It also led to increased food imports by lowering tariffs and exposing local produce to unfair competition. The increased consumption of imported foods and the increases in diet-related diseases are a consequence of this (Deep Ford and Rawlins, 2007; CFNI, 2007).

14.6 Food Sovereignty

According to Dr. Kayano Nwanze, president of the United Nations International Fund for Agricultural Development, "Smallholder farmers supply 90 percent of the food for developing countries and feed one-third of the world." This means that for small developing island states like those of the CARICOM region, any serious effort at enhancing food security must start with increasing local production and improving self-sufficiency. We argue that fundamental to enhancing food security in the region is improving the capacity of local food producers to significantly increase the production and supply of affordable nutritious food produced using environmentally and economically sustainable production systems. In this context we suggest that this is best achieved by adopting elements of a food sovereignty approach (Beckford and Bailey, 2009; Holt-Gemenez, 2006; Schwind, 2005). Food sovereignty has been described as a basis for the revitalization of rural spaces with equitable distribution of resources and small-scale producers having the ability to supply locally grown healthy foods (Windfuhr and Jonsen, 2005). This concept also speaks to the rights of people to determine the source of their foods, to be able to protect and regulate domestic agriculture and trade, and to restrict unfair competition from cheap foreign food imports (Via Campesina, 1996; Institute for Agriculture and Trade Policy, 2003). This incorporates into the discussion the issue of *agency*, with the empowerment of farmers and rural peoples to solve their own problems. Food sovereignty philosophy does not eschew international trade but rather advocates the formulation of trade regulations that serve the interests of local peoples and farmers.

A food sovereignty approach in the Caribbean would prioritize local agriculture by providing farmers with the capacity to produce affordable, healthy, and safe foods while protecting them from unfair competition, which place their livelihoods and regional food security at risk (Schwind, 2005; Kent, 2001). It would enhance food self-sufficiency and reduce the dependence on food imports. Food sovereignty strategy should also increase the participation of farmers and local peoples in agricultural planning and decision-making (Beckford and Bailey, 2009; Stamoulis and Zezza, 2003). We submit that an effective food security strategy will invest in domestic food farmers, empowering them with control over the production and marketing of their crops. This entails a more nuanced trade system that removes barriers to small-scale farmers' viability and does not compromise domestic agriculture (Barker et al., 2009).

Our advocacy of the food sovereignty approach is predicated on the application of a livelihood perspective. Food security and rural livelihoods should be fundamental to agricultural planning in the Caribbean (Beckford and Bailey, 2009). As we showed earlier in this volume, the proportion of the labor force engaged in farming and the contribution of agriculture to GDP in the region are both declining. But they are still significant. In rural areas in particular, farming is still one of the main economic activities for men and women. It is also critical to household food security and nutrition even when farmers practise on a subsistence basis. The survival of many households, access to education especially higher education, home improvements, investments in children's futures, and the general standard of living are dependent on agriculture. It is therefore critical to rural livelihoods and rural development more broadly that the local small-scale farming sector be nurtured and supported.

14.7 Summary

"Agricultural trade and trade policy have been critical to achieving high levels of food security and human development index in many Caribbean states" (Deep Ford and Rawlins, 2007, p. 20). It is neither feasible nor possible for the Caribbean to supply all its food needs. It should be noted that some islands in the region are simply too small to grow all the food they need. For example, countries like Antigua and Barbuda and the Bahamas require much more food than they are able to produce and have very high variance between food exports and imports, but this can be largely attributed to their small size (Deep Ford and Rawlins, 2007). Haiti, on the other hand, has a very large variance as well, but its high dependence on imports is more an indication of its inability to produce domestically (Deep Ford and Rawlins, 2007). Also the uniformity of the climate through the seasons does not create the seasonal variations in weather and climate, which are needed to produce the range of crops required for the region to be self-sufficient in food. This is especially so given the traditional staple foods institutionalized in the culture through slavery and colonialism and the changing food taste in the contemporary Caribbean. But the region has the capacity to be far more self-sufficient and much less dependent on imported food, and subsidized cheap food imports force local farmers into untenable competitive situations (Windfuhr, 2002, 2003; Windfuhr and Jonsen, 2005). This is a global phenomenon because of which all aspects of agriculture suffer, not just crops. In India, for example, imports of dairy products from the EU spelled disaster for family farms and the local dairy industry, and in the Caribbean, importation of pork from the United States had deleterious impacts on the local pork industry (Windfuhr and Jonsen, 2005).

Realistically, food security in the region can be achieved only through a balanced approach between international trade and domestic production (Beckford and Bailey, 2009). The issue is that currently there is too much dependence on food imports. Intraregional food production and trade should also be strengthened to enhance regional food security. The high dependence on imported food in the Caribbean represents potential vulnerability to food insecurity (Beckford, 2012; Deep Ford and Rawlins, 2007). Increasing food insecurity and loss of rural livelihoods are being experienced in some Caribbean countries—a development precipitated by trade liberalization, which resulted in loss of access to markets by traditional exports crops like sugarcane and bananas and increased import of food, with disastrous impact on domestic crops and livestock (Deep Ford and Rawlins, 2007). The authors submit that the observed change in consumption patterns and increases in diet-related diseases in the region are a direct result of trade policies.

Improving food self-sufficiency and reducing dependence on imported food are thus big priorities for local agriculture and food security (Beckford, 2012). This should be framed within the general principles of *food sovereignty* (McMichael, 2009b), or as some prefer, *food democracy* (Lang, 2009a). Food sovereignty speaks to the right of local farmers and peoples to define their own food and agriculture in contrast to having food largely subject to international market forces (Beckford and Bailey, 2009).

CHAPTER 15

Tourism, Local Agriculture, and Food Security in the Caribbean

15.1 Overview

Contemporary trends in the global political economy, particularly as they pertain to trade liberalization and the reshaping of global food chains, have made Caribbean states more aware of the need to reform their agricultural sectors in order to be competitive and to enhance food security. One area of concern is the weak links between the tourism industries of the CARICOM countries and their agricultural sectors, especially the small-scale food-producing sector. Local tourism is booming, while local agriculture stagnates and declines (Thomas-Hope and Jardine-Comrie, 2007; Dodman and Rhiney, 2008). This means that symbiotic relationships between the two sectors are at best weak, and in the specific context of food production and food security, the tourism industry does very little to stimulate local agriculture. This is a long-standing contradiction as evidenced by research in the 1970s and 1980s (see Momsen, 1972; Belisle, 1983, 1984). There have been some suggestions that in the past 20 years or so, the situation has improved (Momsen, 1998; Torres, 2003; Conway, 2004; Rhiney, 2009; Timms, 2006), but recent research indicates that we have a long way to go in getting local foods into hotel kitchens in a significant way (Rhiney, 2009; Ramsee-Singh, 2006; Timms, 2006).

Agriculture remains critical to the Caribbean region and contributes significantly to the GDP, but the Caribbean economy has been moving away from an agricultural orientation to become more service oriented and tourism has become the most lucrative regional product outside of Trinidad and Tobago's oil and natural gas. Tourism is the single major source of income in 16 of the region's 30 countries (Rhiney, 2008), and for the Bahamas, the US Virgin Islands, Cayman Islands, and St. Lucia, it contributes over

60 percent of the GDP (Meyer 2006, p. 2, cited in Rhiney, 2008). The organization of the region's tourism product, however, results in high rates of foreign exchange leakages due in part to a high dependence on imported inputs, especially food (Dodman and Rhiney, 2008; Rhiney, 2008, 2009; Ramsee-Singh, 2006; Timms, 2006). These studies show that tourist resorts and facilities across the region are significant consumers of foreign food stuffs including beverages and that the links between local tourism and local agricultural are generally informal, unplanned, underdeveloped, and tenuous where they do exist. Research shows that the tourism industry provides very limited opportunities for domestic agriculture and that hotel size and ownership are key determinants in this dynamic. For example, large, high-end hotels displayed a greater use of imported food compared to smaller, lower end facilities (Belisle, 1982; Momsen, 1971, 1986; Rhiney, 2009; Timms, 2006). Momsen's (1971) study found that in St. Lucia, large foreign-owned hotels purchased over 70 percent of their food from overseas compared to 33 percent for smaller hotels. While tourism and agriculture have both changed since then, the degree of linkage between the two is still insignificant and far below what is possible.

Creating stronger linkages between the agricultural and tourist sectors can increase the profitability of farming, stimulating agricultural expansion and production and enhancing food security. Stronger linkages between the two sectors will also lead to a more lucrative tourism industry as it would reduce the high rate of foreign exchange leakages, which is currently experienced. It would also have implications for rural livelihoods and development and migration from rural to urban areas. There is currently a high rate of migration from rural farming communities to urban tourist centers as farming becomes less profitable and appealing and people seek better opportunities in centers of tourism.

This chapter discusses the importance of deeper linkages between local agriculture and local tourism, the challenges to creating these linkages, and some strategies for achieving complementary linkages between the two sectors.

15.2 Status of Agriculture and Tourism Linkages in the Caribbean

Tourism is an increasingly important source of income for many small island developing states (SIDS) like those in the Caribbean (Dodman and Rhiney, 2008). In the Caribbean tourism is the fastest growing and most dynamic sector of regional economies. Income from the industry is as high as 60 percent of the GDP and an increasing number of islands depend disproportionately on the industry. The importance of the industry in the Caribbean is under-scored when one considers that for 16 of the 30 countries in the region, it is the highest earner of foreign exchange and for some it contributes over 60 percent of the GDP—Cayman Islands, St. Lucia, the Bahamas, and the US Virgin Islands are prime examples (Meyer, 2006). It has, however, failed to solve the economic problems in the region, which research spanning nearly four decades indicates does not maximize the benefits of the industry due to very high rates of foreign exchange leakages precipitated mainly by the heavy outflow of expenditure for imported food (Rhiney, 2010; Timms, 2006; Haley, 2005; Belisle, 1983; Momsen, 1972). The sector does provide a hard-to-quantify, but significant, indirect benefits to island economies and employs substantial numbers of people, but much of this employment is low paid and seasonal.

Caribbean economies evolved as monoculture enterprises for export, while importing virtually all of their basic needs. After emancipation, a domes-tic agricultural sector emerged out of the growth of a local peasantry, which produced food for the local market. Today these two sectors oper-ate largely independent of each other—in some ways even in competition with each other—in an unequal relationship characterized by unequal access to resources and institutional support. Domestic agriculture in the region has declined in recent times, with all the countries except Guyana and Belize now being net importers of food. According to Timms (2006), the lack of link-ages between agriculture and other sectors of the economy including tourism is a major reason for the underperformance of the domestic food sector. The result is high rates of external leakages of foreign exchange as export earnings are spent to finance foreign food imports.

Efforts to improve linkages between agriculture and other sectors first found theoretical ground in the work of Nobel Prize laureate Sir Arthur Lewis in the 1950s. Lewis theorized that growth in other areas of the economy would increase demand for agricultural products (and stimulate agricultural expansion) (Lewis, 1954, 1955, 1958). This was an economically sound argu-ment and it provided a framework for development policy in the region. Unfortunately development in the region did not exactly follow Lewis's the-ories and the linkages he envisioned did not materialize (Timms, 2008). Manufacturing, for example, expanded without establishing any symbiotic relationships with agriculture and other sectors of the economy. This has become a characteristic of Caribbean development.

This lack of intersectoral linkages in Caribbean economies is epitomized by the case of tourism and local agriculture. A major irony of Caribbean agri-culture is what might aptly be described as the estranged relationship between the region's world famous tourist industry and its local agriculture. This is not a new phenomenon as indicated by research from the 1970s and 1980s lamenting the limited benefits the small-scale domestic food sector enjoyed from the tourism industry (Momsen, 1972; Belisle, 1983, 1984). In more recent work it has been suggested that several changes, including more openness toward serving local cuisine in resort facilities and globalization of food consumption habits and the desire of tourist to eat local foods, could serve to strengthen the link between local agriculture and tourism (Momsen, 1986, 1998; Torres, 2003; Conway, 2004; Timms, 2006; Rhiney, 2009). However, recent research into the role of tourism in local food supply chain suggests that there are still considerable problems for farmers forging stronger ties with the local tourism industry (Timms, 2006; Rhiney, 2009). The tourism sector is an area where the potential for creative use of more local foods in the cuisine could be harnessed and successfully promoted (Beckford et al., 2011). The extent to which locally grown foods and traditional foods are used in the hotel kitchens requires research but indications are that use is limited.

Momsen (1972) conducted research in St. Lucia and found that 70 percent of the value of food was imported. Furthermore, 70 percent of large hotels complained about uncertainty of local supply as a major factor in their low use of local foods. Momsen found a discrepancy in reliance on imported food between the four largest hotels that used almost no local foods and smaller, locally owned facilities that used mainly local food. In a follow-up study in 1985, Momsen found that the percentage of imported food used by the largest hotels had fallen significantly to 58 percent by 1983 (Momsen, 1986). This finding was supported by Timms (2006), who found that more supply chain avenues had opened up and that the larger facilities were doing a better job using local farm produce. Belisle (1984, 1983) confirmed this finding in a study comparing the use of local and imported food by hotels in Jamaica from 1979 to 1980. Ramsee-Singh (2006) indicates that Jamaica experiences a 50 percent rate of foreign exchange leakage. This is due to many factors, chief among them being the high import content, which characterizes the industry (Belisle, 1984; Pattullo, 2005; Jayawardena and Ramjee-Singh, 2003; Hayle, 2005; Ramjee-Singh, 2006, Timms, 2006; Dodman and Rhiney, 2008), repatriation of profits due to the large foreign ownership, and high expenditures on foreign advertising (Jayawardena and Ramjee-Singh, 2003). The high percentage of imported food is a major source of leakage. Rhiney (2009) found that hotels in Negril, Jamaica, consume a high amount of imported food. The Jamaica Tourist Board (2005) revealed that perhaps 70 to 90 percent of food used in small hotels is local but only about 40 percent of the food consumed in the large properties is local (Pennicook, 2006). Other factors that could explain this fact is that large hotels deal in international cuisine and tend to hire foreign head chefs.

15.3 Benefits of Improved Links between Local Agriculture and Tourism

Tourism has fallen short of its potential to lead to regional development due to weak external linkages with other sectors of the economy such as agriculture as demonstrated by high import food bill for the industry and the heavy outflow of foreign exchange earnings (Brohman, 1996). As Timms (2006) opines there are obvious benefits to be derived from improved synergy between tourism and other sectors. These could include an increase in the marketing of local agricultural produce. Creation of reliable markets would stimulate expanded production and create demand for agroprocessing infrastructure, which has been neglected in favor of export agriculture.

Momsen (1998) suggests that there is potential for seasonally complementary labor demands between agriculture and tourism, which would lead to the creation of jobs and reduction in underemployment and ultimately raising annual incomes. The problem with this argument is that farmers and agricultural laborers typically have very low educational attainment and most of them would perhaps be unemployable in the hospitality industry.

Greater linkages between agriculture and tourism would create increased demand for varied and high quality foods that should serve as an impetus for farmers to diversify and expand production. Greater linkages between the two sectors could also create greater access to credit for local producers. Lack of access to credit has been a historical constraint to domestic agriculture. As we discuss later in this chapter successful farmers' cooperatives that supply a large hotel chain in Jamaica and St Lucia have benefitted from financing from foreign sources. Their link to the tourism industry was key to their success. It has been argued that the necessary linkages rather than being planned for were left to develop spontaneously (Torres, 2003). It has also been suggested that the potential to stimulate linkages between agriculture and mass scale tourism is little to none (Pattullo, 2005).

15.4 Obstacles to Improved Linkages between Local Agriculture and Tourism in the Caribbean

The relationship between local agriculture and local tourism is characterized by disorganized and weak linkages between the two sectors. This is true of all the countries in the region but is perhaps most striking in the larger islands, especially Jamaica—the largest island in the CARICOM though not the largest country.

Research from Jamaica shows a complex set of relationships between the two sectors with significant challenges on all fronts (Rhiney, 2010). From the perspectives of hotels especially the large all inclusive facilities there is a lack of communication with suppliers, increasingly high demand for both higher quality and larger quantity of food on a consistent basis, and their preference for the cheaper, easy to prepare food items which are most easily sourced overseas. Farmers and suppliers for their part are unable to meet the qualitative and quantitative requirements of hotels on a consistent basis year round. This is problematic for hotels and is a major obstacle to the development of any sustainable links between the two sectors. It is easier and more efficient for large hotels to use imported frozen semi-processed and processed foods than fresh local produce. Not only are these easier to use but they also have a longer shelf life than the perishable fresh local produce. There are explanations that local fruits and vegetables are often infested by insects and preparation times in hotel kitchens are longer as produce must be carefully washed and sorted to remove insects and their residue. In contrast foreign foods are usually much cleaner and hence quality is assumed to be better. Of course if analyzed fully it becomes clear that the foreign produce is cleaner because of the liberal use of preventative agrochemicals, which raises questions about food safety and human health.

The agriculture sector faces a huge challenge to guarantee sufficient and reliable supplies of high quality competitively priced produce. Important issues in this regard include-poorly organized production and distribution systems, infrastructure deficiencies, competition from imported food, climatic hazards, and lack of access to resources and services. All of these add to an unfavorable agro-ecological environment.

Supply of food to hotels is dominated by a large and growing group of intermediaries who trade in the resale of food. Rhiney (2010) explains that under this arrangement a hotel has an agreement with a purveyor to supply food at a set price. The purveyor then goes out and sources the food at a lower price in order to be profitable. This is a complex, unplanned informal distribution system between farmers and purveyors. Food is supplied by the farmer on credit to the purveyor who pays after the sale has been made to the hotel or restaurant. Farmers have reported being cheated by intermediaries and that they often have to wait long periods to receive payment for their produce. Some say they that in the past they have on occasion not been paid at all. Farmers are at left without recourse as the loose and informal nature of the business mean that typically they have no written contract or binding agreement that can be legally enforced.

The system works to keep farmers and hoteliers apart as well. They never come into contact or communicate with each other. They have no information or knowledge about the others needs and requirements and thus are not responsive to each other. In reality both are to some extent at the mercy of the purveyor but the hotel has far more resources and power to protect itself.

Natural disasters like hurricanes play a role in the divide between local agriculture and the tourist industry as well. Rhiney (2009) shows how after major hurricanes hotels typically use even more imported food as the local food production sector is disrupted. Local farmers can experience difficulty reintegrating into the supply chain as the recovery process can be protracted. Emergency relief to farmers in the aftermath of disasters is invariably slow, inadequate, and characterized by political patronage and clientelism.

As is the case with food production and distribution in general inefficiencies in local cropping systems also impact the development of stronger linkages between local agriculture and tourism. There is lack of coordination of production between farmers and their decision-making is not based on reliable market information. Farmers basically plant crops based on traditional combinations and timing and farmers in an area more or less plant the same crops at the same time. Seasonal gluts and depressed prices often exist for some crops with scarcity and high prices for other crops at the same time.

Infrastructural and technological deficiencies also plague local agriculture and affect the ability of farmers to benefit from the tourism industry. For example, lack of knowledge and expertise about sorting and grading makes it difficult for quality control and this affects prices and hence income. Local farmers do not sell produce according to grade and so don't benefit from charging higher prices for higher quality produce. Lack of knowledge about proper food storage and lack of storage facilities are also major issues. This prevents farmers from controlling production and distribution cycles and aligning their activities better with the tourism industry. Often production is high when there is little or no demand in the tourism sector or low when the demand is high. Thus there is no capacity within the local food production sector to guarantee consistency of supply to hotels and restaurants.

The easy access available to large hotels to procure competitively priced food overseas is a major obstacle to the development of deeper linkages between local agriculture and tourism. The local tourism industry can exist independently of the local agricultural sector because of alternate sources of cheap food. Deregulation and trade liberalism policies have opened up Caribbean markets to cheap foreign foods produced under subsidized arrangement chiefly from the United States. Local farmers are unable to compete with these cheap foods. Purveyors often purchase these supplies in their arrangement with hotels since they are cheaper than local produce. The availability of this cheap source of food is a disincentive to hoteliers, restaurateurs and purveyors to buy local food and is a huge disincentive for local farmers to as well.

There are many significant obstacles to deeper links between tourism and local agriculture in the Caribbean but according to Rhiney (2010) these problems are not intractable. Macro scale factors are important. The policies of governments have not encouraged or facilitated connections between the two sectors. Instead they have developed in isolation of each other and the existence of linkages as they are is in spite of rather because of government intervention and planning.

With intensive and highly successful efforts to increase revenues from the tourism industry by increasing visitor numbers and mobilizing foreign direct investments the tourism industry can be expected to grow even more. Unless greater linkages are created between agriculture and tourism this growth will simply lead to increases in food importation. The last decade has seen numerous initiatives aimed at stimulating food production and enhancing food security in the Caribbean. None of these has been geared specifically towards fostering deeper linkages between local agriculture and local tourism. In March 2013 Jamaica's Minister of Agriculture announced that consultations were being held between industry leaders that would lead to a policy for developing inter-sectoral linkages between tourism and other sectors of the economy.

15.5 Agriculture and Tourism: Food Supply Networks in the Caribbean

Insights into the weak linkages between tourism and local agriculture can be gleaned by studying strengths and weaknesses of the local food supply networks. Rhiney (2009) and Timms (2006) in their recent work on Jamaica and St. Lucia respectively have suggested that supply chains and networks between farmers and hotels should be included in the analysis of tourism and agriculture linkages. Rhiney argues that to examine the factors shaping agriculture-tourism linkages it is important to explore the constraints faced by the local food network in producing reliable and high quality food supplies on a consistent basis. Timms (2006) argues that linkages should be redefined as relations and include issues like information flow and communication between farmers and hotels.

Momsen (1972) identified seven different types of suppliers of food to hotels in St. Lucia—contract, itinerant hawker, market, marketing board, agent, supermarket, and own estate. Timms (2006) also in St. Lucia found nine different distribution types—formal contract with individual producers, cooperatives, informal contract with individual producers, wholesaler, agent/hawker, informal contracts with vendors, local markets, supermarkets, and self-supply. This suggested that links had expanded since Momsen's work.

In his study of the links between local agriculture and the tourism industry in Negril, Jamaica, Rhiney (2008, 2009) identified several networks of providers in the hotel food supply chain. These were local purveyors, wholesalers, retailers and supermarkets, farmers, farmers' cooperatives, and local produce markets. By examining the two main players in the supply chain network the problems and potential in relation to the forging of more meaningful linkages between local agriculture and tourism are highlighted.

Farmers generally depend heavily on purveyors who provide an important avenue for them to market their crops. For some this represent their only real outlet. But the purveyor system is fraught with problems and has not been effective in providing a reliable supply of adequate quantities of good quality food to the tourism sector. There are many reasons for this. Purveyors typically use open back pick-up truck for transporting food that is usually poorly packed. Refrigerated and cold storage vehicles are rare. With the long distances traveled fresh produce is not always so fresh when they are delivered to hotels. Spoilage and bruising are common especially in the case of some fruits and vegetables. Hotels often express dissatisfaction with purveyors who they accuse of being unprofessional as they are often late and are inconsistent with price and quality (Rhiney, 2009). Many purveyors also supply imported food to hoteliers who do not seem to have a good sense of the source of the food delivered through this avenue.

Farmers are dissatisfied with the purveyor system as well. Rhiney (2009) documented deep distrust on the part of farmers almost all of who have bad experiences with the purveyors. A major source of angst is the length of time it takes purveyors to pay them after taking their produce on credit. It is not uncommon for farmers to not be paid at all. Farmers also feel cheated by purveyors in the price they are offered for their produce. There is a view that these intermediaries buy at very low prices and then sell much higher to maximize profits. Due to the dynamics of the demand and supply many farmers feel compelled to accept low prices from the purveyors. The informal nature of the purveyor system makes it disorganized and unreliable.

In contrast to the purveyor system both Timms (2006) in research from St. Lucia and Rhiney (2009) in his Jamaica study found evidence of the potential of cooperatives in strengthening linkages between local agriculture and the local tourism industry. Rhiney (2009) concluded that cooperatives represented one of the best strategies for enhancing agriculture and tourism linkages in Jamaica. Unlike the purveyor system cooperatives are formal arrangements. They allow farmers to pool resources and accomplish things that they could not otherwise do as individuals. One cooperative in Jamaica had allowed farmers to mobilize resources such as accessing loans, gaining access to land and training and developing partnerships with foreign organizations including USAID, CIDA, UNDP, and IICA (Rhiney, 2008). In other words this cooperative has managed to benefit from the forces of globalization. Through accessing international financial aid amounting to over J$6 million this cooperative has modernized and expanded its operation with the construction of packing houses, cold storage, nurseries, greenhouses, and irrigation (Rhiney, 2009).

The cooperative system has facilitated greater communication between farmers and hotels. By eradicating intermediaries, better two-way information flow opens up leading to better coordination of demand and supply. Cooperatives have been more successful in providing consistent supply because members engage in staggered planting and no doubt because they are able to store produce and engage in better post harvest management practices. Because the costs of production are shared and the risks spread, they also experience benefits derived from greater economies of scale (Rhiney, 2009). The cooperative system described by Rhiney came out of the Sandals Farmers' Program that started in 1996 with ten local farmers supplying the Sandals chain of hotels with food. The program was expanded to St. Lucia, where Sandals operated a number of facilities and it has brought significant benefits to farmers involved (Timms, 2006; Tapper and Font, 2004; Abdool and Carey, 2004).

Despite their big upsides, cooperatives are not without challenges. They too suffer from competition from cheap imported food, are vulnerable to the vagaries of the Caribbean weather, and experience problems with local hotels that are not always consistent with their purchasing patterns. Timms (2006) identified one of the bottlenecks in the agriculture tourism relationship as the desire of hotels to retain somewhat flexibility in their food supply sources which lead to them avoiding cooperatives and binding contracts.

15.6 Summary

Caribbean development is characterized by a lack of intersectoral linkages. This has led to vital sectors of the economy developing in isolation of each other and therefore losing out on the reciprocal benefits that accrue from inter- and intrasectoral synergies. One area in need of special attention is the link between the tourism industries of the CARICOM countries and their agricultural sectors especially the small-scale food producing sector. Local tourism is expanding while local agriculture is in decline and dependence on imported food is growing (Thomas-Hope and Jardine-Comrie, 2007; Dodman and Rhiney, 2008). This is nothing new as research in the 1970s and 1980s indicate (see Momsen, 1972, 1986; Belisle, 1983, 1984). The literature suggests that there is a slight improvement in the links between local agriculture and local tourism in some Caribbean islands, example St. Lucia and Barbados. There have been some suggestions that in the past 20 years or so there has been more of a willingness on the part of visitors to eat local cuisine, a maturing of the regional tourism product and increasing assistance to local farmers by governments (Momsen, 1998; Torres, 2003; Conway, 2004; Rhiney, 2009). Timms (2006) found that large hotels in St Lucia had indeed begun to use more local food and suppliers confirming Momsen's findings (Momsen, 1986). However, this is not always the case as recent research indicates that in some islands we have a long way to go in getting local foods into hotel kitchens in a significant way (Rhiney, 2009). Some researchers are unconvinced about the ability of the two sectors to forge greater ties (Pattullo, 2005) as leakage rates are high and the region remains tied to foreign food sources. Dodman (2004) was hopeful that the global shift away from traditional models of mass tourism to small-scale alternative forms of tourism, is leading to new niche markets and these may stimulate increase demand for local food among visitors (Rhiney, 2009).

Farmers have argued that there is need for a more proactive state role in the marketing and distribution of domestic food crops to enhance viability by providing stable markets and fair prices. The constraints to food production are essentially structural issues in the organization of production and marketing and are critical components of the weak links between agriculture and tourism. These constraints function to impair the ability of the local food sector to meet the needs of the tourism industry. Until these are addressed links between the two will remain weak and disorganized. It is imperative that governments take steps to increase the capacity of local farmers to supply high quality competitively priced fresh produce. This requires action on both the production and distribution aspects of the issue. Government support should also encourage and educate farmers about establishing farming cooperatives for the production and marketing of food.

Promotion of cooperatives or production and marketing organizations (PMOs) can be an effective strategy for forging more substantial and significant links between local agriculture and local tourism. Not much has been written about the co-operatives as alternatives to traditional farming systems in the Caribbean but recent work, for example, Timms (2006) in research in St. Lucia have identified some qualified potential to stimulate local production and foster intersectoral linkages between agriculture and tourism. Rhiney (2008, 2009) have also documented encouraging cases in research on Jamaica. Beckford et al. (2011) in calling for greater linkages between tourism and the yam industry in Jamaica suggested that yam farmers would be better served by establishing cooperatives to cultivate and market their produce rather than acting individually.

Access to cheap foreign foods by hotels in the region is a major obstacle to greater linkages between the tourism industry and local agriculture. Caribbean small-scale food farmers are unable to compete with cheap imported food, which was produced under the protection of attractive subsidies but the deregulation of the local economies has flooded local markets with foreign foods. The ability of small-scale food producers in the Caribbean to develop meaningful links with the region's tourism industry is dependent upon their ability to function in an increasingly open market economy characterized by the proliferation of cheap imported food and lack of state support. In the absence of state encouragement and support, they have not yet shown the ability to do so. Enormous opportunities exist particularly among large, all-inclusive, hotels to stimulate local food production in the Caribbean. But this is an hitherto unrealized potential.

PART V

Synthesis

CHAPTER 16

Improving Food Security in the Caribbean: Building Capacity in Local Small-Scale Farming Systems

16.1 Overview

The food security challenges of the Caribbean, and the developing world more generally, cannot be solved by dependence upon food imports. Although globally there is enough food to feed everyone, world hunger remains a dire problem. Our view is that to achieve real food security, developing countries must become more food self-sufficient by increasing productivity, diversifying and expanding the range of crops with a focus on maximizing the use of traditional foods, reducing postharvest losses, improving the marketing and distribution of farm produce, promoting urban agriculture, and increasing women's participation in the food security endeavor. Our position is framed within the general principles of food sovereignty (McMichael, 2009b), or as some prefer, food democracy (Lang, 2009a). Food sovereignty speaks to the right of local farmers and peoples to exert more control over food and agriculture (Beckford and Bailey, 2009). Windfuhr and Jonsen (2005) described food sovereignty as a platform for rural revitalization at the global level based on equitable distribution of resources, farmers having control over resources, and the ability to supply healthy, local food. A food sovereignty approach advocates the right of people to be able to protect and regulate domestic agriculture and trade in order to achieve sustainable development goals: to determine the extent to which they want to be self-reliant and to restrict dumping of products in their markets. This incorporates into the discussion the issue of agency, that is, the empowerment of farmers and rural peoples to solve their own problems. Such approaches would specifically draw on local agro-ecological knowledge and wisdom of elders.

Over 90 percent of food in developing countries is produced by smallholder farmers, which suggests that Caribbean small-scale farmers must play a greater role in enhancing food security and food self-sufficiency in the region. Because of this, building the capacity of small-scale food producers to increase agricultural output becomes critical. An important issue is reimagining the role of women in the production and marketing of food. This is important to national food security, but it is critical to community and household food security and nutrition. To be effective and sustainable, food security strategy in the CARICOM must create the conditions for females to improve their own food security and their families, improve nutrition, and achieve greater economic independence.

Also important is an emphasis on activities to strengthen local food production and distribution systems and to increase the capacity of farmers to produce adequate quantities of nutritious and affordable food through sustainable systems and practices and to increase income and improve livelihoods. As part of the strategy to accomplish this, initiatives should address the role of women in the production and marketing of farm produce and nutrition. The overall goal should be an integrated and comprehensive strategy aimed at building the capacity of small-scale food producers in the region to increase productivity and improve livelihoods and income through gender-responsive sustainable agricultural technologies and practices.

16.2   Enhancing Food Security through Increased Food Self-Sufficiency

The empowerment of local food producers in the Caribbean should focus on activities aimed at strengthening local food production and distribution systems and increasing the capacity of farmers to increase food production through sustainable systems and practices and to increase income and improve livelihoods. The overall goal should be the development and implementation of an integrated and comprehensive strategy aimed at building the capacity of small-scale food producers in the region to increase productivity and improve livelihoods and income through gender responsive-sustainable agricultural technologies and practices. In the next section we suggest a number of specific areas that research tells us need to be addressed to enhance the capacity of local farmers to produce large amount of food, which is more diverse, better in quality, and affordable.

16.2.1   Enhancing Farming Expertise

Our research leads us to believe that increasing farmers' knowledge and understanding and their ability to apply this knowledge and understanding is a fundamental issue in capacity building and empowerment of local food producers. Research among small-scale farmers in the Caribbean consistently points to a wealth of traditional or local knowledge based on intergenerational knowledge and experience (Beckford and Barker, 2007; Campbell, 2011; Hills and Iton, 1983). This has served farmers well over the years and is largely responsible for the success and survival of many small-scale food producers who have been forced to survive without any significant institutional support (Beckford and Bailey, 2007). The education of farmers, we envisage, is a structured and systematic program of practical information dissemination based on evidence-based identification of the gaps in farmers' knowledge and priorities for action. This dissemination should be done through agricultural extension services, farmer field schools, and on-farm adaptive research based on participatory and collaborative principles. Based on recent research from Jamaica, priority areas identified by farmers include postharvest storage of crops, irrigation, pest management and control, marketing and distribution of crops, organic farming, record keeping, grading of fresh produce, soil management, and adaptation and coping strategies with regard to meteorological hazards, which are ubiquitous to the Caribbean (Campbell and Beckford, 2009; McGregor et al., 2009; Beckford and Bailey, 2009; Rhiney, 2009).

16.2.2 Postharvest Management

Postharvest management of food among small-scale farmers in the Caribbean requires urgent attention. Proper storage facilities for farmers are lacking as are knowledge and experience of general advanced postharvest management of farm produce. Few farmers keep good records, and so the volume of postharvest crop loss is hard to quantify, but evidence gathered from field research suggests they are significant. Crops most affected include fruits, vegetables, condiments, and peas and beans. There is little or no capacity to store produce during periods of oversupply and to release food to the market at different times, thereby sustaining income over longer periods. The economic viability of domestic food cultivation can be significantly enhanced by reduction in food loss due to damage and decay. This would allow farmers to increase their production, as they would know that they would have a much longer window for marketing their produce as well as regulating market supply (Beckford et al., 2011; Rhiney, 2010). There are examples from places with similar experiences that offer lessons on solutions that are often simple and inexpensive. In Tamil Nadu for example, post harvest loss of potatoes was significant as they were stored in mounds and the ones in the middle and at the bottom rotted quickly due to trapped moisture (Bechard, 2010). This was mitigated by inserting plastic pipes with holes drilled in them into the piles, which facilitated air exchange and circulation and slowed down the decay process (Bechard, 2010). Also in Tamil Nadu, postharvest losses of chili peppers have been reduced by 95 percent using simple solar tunnel dryers (Bechard, 2010).

16.2.3 Grading of Fresh Produce

Education and training in the proper cleaning, sorting, and grading of fresh produce—especially for the export market—is also needed. This would allow farmers to market produce differently based on quality. Training in sustainable pest management is also very important. There needs to be more efforts at promoting non-chemical pest control protocols among small-scale food producers. The promotion of organic fruit and vegetable production among small-scale producers is just as important. This would cut the use of chemical fertilizers and raise farm incomes through higher prices while supplying healthy and nutritious produce to the market.

16.2.4 Natural Hazard Mitigation

Education about natural hazard mitigation is also needed. The region is prone to a host of such hazards with hurricanes getting most of the attention, but droughts, floods, and landslides are all features of Caribbean life. In the past ten years there has been billions of dollars in agricultural losses due to hurricanes and tropical storms. In Jamaica, for example, a cycle of storms caused J$993 million of losses in 2005 alone (McGregor et al., 2009). As we discussed in Chapter 10, hazard impact is difficult to quantify as there are multiple events that might have to be accounted for in any given year. For example, in 2005 in Jamaica severe droughts were experienced in the first four months followed by Hurricanes Dennis and Emily and tropical storm Wilma between August and September (McGregor et al., 2009). Farmers need to be aware of what they can do before and after extreme events to mitigate losses. For example, how can plants already in the ground be safely removed until after a hurricane? What crops can be harvested early and how should they be stored to increase their chance of survival?

16.2.5 Record Keeping

Small-scale food producers in Jamaica and the rest of the CARICOM region keep very poor records of their farming operations. The vast majority do not keep formal records. This makes it difficult to accurately analyze their operations and identify reliable solutions to their problems. Their operations are typically very informal with little application of principles of business. Even for the multitude of farmers for whom crop cultivation is their only source of income farming appears to be more of a way of life than a business. Record keeping would ensure accurate applications of inputs and facilitate better planning. Farmers also need education about the marketing and distribution of produce.

16.2.6 Irrigation, Soil Management, and Disease and Pest Control

There is also a need to upgrade farmers' knowledge in the areas of irrigation and drought management, soil management, and pest control. Small-scale farmers possess a vast wealth of traditional knowledge and skills related to these areas, but with the aging of the farming population, these traditional methods are at a risk of being lost. This knowledge is often local and unique to different farming communities and should therefore be made known to the general farming community so that farmers with the help of agricultural extension services could determine their applicability to their needs and agro-ecological contexts. Initiatives to combine Western scientific knowledge with farmers' ethno-scientific knowledge could bring benefits to the agricultural sector through increased sustainability in agronomic practices, reduction in crop losses, and increases in yields.

16.3 Rediscovery of Local Foods

The problem of food security in the Caribbean requires local solutions that should include the discovery and rediscovery of local or traditional foods. Food security in the Caribbean is being undermined by changes in tastes and diet to North American influences as people become estranged from their local foods and consumption habits (Beckford, 2011; CFNI, 2007). This is not unique to the Caribbean. In Lebanon for example, it has been found that food security is compromised as local people transition from a traditional, diverse Mediterranean diet to Western-style diets deficient in micronutrients and heavy on white flour, corn, sugar, and vegetable oils, which are not as nutritious as local olive oil (Boothroyd, 2010, p. 15). This has resulted in a high incidence of high blood pressure and high cholesterol among people 40–60 years old.

Many traditional foods and wild and edible plants have lost their place in local diets. There is a need for reintroducing some of these plants and foods back into the local diets. Research and education is needed about local edible wild plants. These should be documented highlighting their uses, preparation, and nutritional and health benefits. A study in Lebanon funded by Canada's International Development Research Center (IDRC) found that villagers who regularly used wild edible plants and kept gardens enjoyed greater food security and better health than those who did not (Boothroyd, 2010). The nutritional value of over 40 edible plants was studied and nutritious affordable dishes they made were identified. There should also be efforts to rediscover traditional medicinal plants and home remedies that were once widely used throughout the Caribbean. These should be catalogued and documented with their properties and uses identified and they should be included in education curricula at all levels as well being available on the internet and in libraries.

16.4   Increasing Women's Participation in Food Production

Sustainable food security in the Caribbean requires the effective participation of women in food production. This is significant in the context of the dimensions of availability, access and nutritious foods and the implications for overall household food security. There are many commercial female farmers in the Caribbean but women are mainly involved in the marketing and distribution of food. Greater participation of women in food production could be an effective strategy for addressing food security at the household level. This can be done through a Backyard Garden Project in which women receive training in growing organic foods especially fruits and vegetables mainly for home consumption. The aim would be to increase supplies of nutritious foods for their households. Women should also receive training in food handling and preparation to maximize the nutritional value of their families' meals.

An interesting aspect of the IDRC study discussed earlier that holds lessons for the Caribbean is the development of a communal "Healthy Kitchen" by women in three villages. The project centered on the preparation of traditional dishes using wild plants and other produce. Training in commercial food preparation and marketing was provided and the women sold their produce in local markets, catered at weddings and other events and showcased their Healthy Kitchen cuisine. This is the kind of approach that may be necessary to bring Caribbean populations back to local and traditional foods. It is different from conventional eat local campaigns conducted through the media which have been largely unsuccessful due to the top down approach and lack of grassroots community and household engagement.

16.5   Community and Household Agroprocessing

Most of the food produced in the Caribbean is sold as fresh produce. An important component of food security and women's participation should be initiatives to promote increases in agroprocessing at the household or community levels. Agroprocessing would drastically reduce postharvest losses, preserve food, and add value, thus increasing farm incomes. Cottage industries based on locally produced fresh farm produce should be promoted and supported. In this regard there should be efforts to establish properly constituted cooperatives but household-level industries should also be pursued. Women could also play an instrumental role here as the history of cottage industries in the region suggests that they have always taken a leadership role. Again there will be a need for training and ongoing learning in areas like food processing, business management, marketing and distribution and accounting.

16.6 Distribution and Marketing of Fresh Produce

This is an area requiring urgent attention. The marketing and distribution of domestic food crops is done through various informal commercial activities. The primary strategy is where farmers sell their produce to people mainly women who sell in produce markets across the country. The marketing and distribution of domestic foods is a major obstacle to production in the Caribbean. There is no regulated system in place and small-scale farmers are basically left to their own devices in marketing farm produce locally. One response to this has been the recent establishment of farmers' markets where farmers can take their produce directly to the market place and sell consumers.

A key concern is the state of linkages between local tourism and local agriculture sectors especially the small-scale food producing sector. Paradoxically local tourism is booming while local agriculture stagnates and declines (Thomas-Hope and Jardine-Comrie, 2007; Dodman and Rhiney, 2008) a long-standing contradiction as evidenced by research in the 1970s and 1980s (See Momsen, 1972; Belisle, 1983, 1984). While changes have occurred linkages between tourism and agriculture remain weak. Farmers have argued that there is need for a more proactive State role in the marketing and distribution of domestic food crops to enhance viability by providing stable markets and fair prices. Government support should also encourage and educate farmers about establishing farming cooperatives for the production and marketing of food.

16.7 Urban Agriculture

Like most of the developing world, the Caribbean is characterized by increasing rate of urbanism. This is due to both natural increase and rural to urban migration precipitated by rural marginalization and underdevelopment. In Chapter 9 we discussed the potential of urban agriculture to play a vital role in food security especially household nutrition. We also discussed the lack of an urban agriculture strategy in the Caribbean as a whole. Urban agriculture takes place in the cities of the region but the extent to which it is practiced is not quite known as there has not been enough research on the subject. There is, however, no coherent strategy and where urban agriculture exists it is haphazard, unregulated and unsupported and based on personal ingenuity of participants. In 2010 a Regional Food and Nutrition Strategy Policy was developed for the Caribbean. This policy makes no mention of urban agriculture at all despite its growing importance globally. Much closer to home there is the world standard in urban agriculture policy and program in Cuba. CARICOM governments should be modeling Cuba's commitment to and support of urban agriculture as a central component of regional food security strategy.

16.8 Greater Focus on Nutrition and Access to Food

Food insecurity in the Caribbean (with the exception of Haiti) tends to be under stated. It is often construed simply in terms of availability and complacency exists because again with the exception of Haiti, the dramatic and sensationalized incidents of hunger often seen in parts of Sub-Saharan Africa and Asia are largely unknown in the region. Discussions about food security in the region tend to be overly centered on food availability and other dimensions of the issues such as nutrition and entitlements or access to food are neglected. Here we call for a greater focus on the nutritional and access dimensions of food security in policy, practice and research in the Caribbean. A 2007 study found that in the CARICOM region, while contraction in local food production is cause for concern compromises to food security are delineated not so much by food availability but more importantly by inadequate access to food and dietary patterns which has deleterious effects on nutritional status (Caribbean Food and Nutrition Institute [CFNI], 2007). In many Caribbean countries estrangement from traditional diets have led to increasing rates of nutrition related deceases such as hypertension and diabetes as obesity rates increase (Technical Working Group on RFNSP, 2010). High levels of unemployment have led to increasing poverty and widening gaps in income distribution making it difficult for increasing numbers of Caribbean people to meet their food needs in the context of rising food prices. The food security issue is therefore multi-dimensional and requires a more holistic approach in the Caribbean.

16.9 Local/Traditional Knowledge

Honoring and supporting local or traditional agricultural knowledge and practices should be part of any strategy to enhance food security in the Caribbean. The validity and value of traditional skills and knowledge in small-scale agricultural systems have long been recognized (see Chambers, 1983; Chambers et al., 1989; Richards, 1985; Hills and Iton, 1983). In the Caribbean local knowledge has evolved as a sound adaptation to local realities. It is a necessary adaptation to resource constraints in a challenging agro-ecological, cultural, economic and political milieu rife with risks and uncertainties (Beckford and Barker, 2007). Successive generations of small-scale food farmers in the region have successfully used traditional agronomic practices related to soil management and conservation, disease and pest control, moisture retention, crop combinations and cropping systems among other things. Small-scale food farmers in the Caribbean have traditionally operated in an environment where they are fending for themselves and depending on their own ingenuity to make a livelihood. Food security strategy should seriously consider the role that local knowledge plays in sustainable food production.

16.10 Praedial Larceny

Praedial larceny involving crops, livestock, and fish is now recognized at the highest level as a major constraint to sustainable agriculture in the Caribbean (CARICOM View, 2011). Eighty two percent of farmers say that they have been affected by praedial larceny at some point. It is estimated that regional losses amount to over US$ 350 million annually (CARICOM View, 2011). In Belize it is estimated that thieves cost farmers and fishermen over US$ 300,000 annually while in Jamaica the figure was US$ 55 million (CARICOM View, 2011, citing various sources). In St. Lucia US$ 400,000 is being spent on a pilot program on praedial larceny prevention. There is evidence that the problem is getting worse and that thieves are more violent and more organized. More research is needed to enhance our understanding of this problem and its impact on food availability and security in the Caribbean. What is evident, however, is that praedial larceny is a huge disincentive to farmers. The problem needs to be addressed as a matter of urgency.

16.11 Summary

This discussion has demonstrated that food security is an area requiring urgency in regional development. The region has made some strides in addressing food security concerns but there is still a great deal of work to be done from the standpoint of policy but also at the grassroots level. Erwin Larocque, CARICOM's Secretariat's assistant secretary-general for regional trade and Economic Integration underscored this in highlighting two pressing regional food security issues. First, he made reference to the impact of international developments on the Caribbean's ability to be self-sufficient in food production and internationally competitive to afford necessary imports. Second, he stressed the need to address food security issues in the context of the Millennium Development Goals. He stressed that the region's dependence on imported food made it vulnerable and noted the climb in diet related diseases. In the Anglophone Caribbean only Belize and Guyana are currently net importers of food and the gap between food imports and agricultural exports for other territories is over 60 percent—over 80 percent for Antigua and St. Kitts/Nevis (Deep Ford and Rawlins cited by Barker et al., 2009).

Improving food self-sufficiency and reducing dependence on imported food are thus big priorities for local agriculture and food security. Based on previous research with small-scale farmers in Caribbean this paper identified a number of specific suggestions for addressing food security in the region. These may be summarized as follows: (i) increasing farmer expertise to in areas such as postharvest storage of crops, irrigation, pest management and control, marketing and distribution of crops, organic farming, record keeping, grading of fresh produce, soil management and adaptation, and coping strategies with regard to meteorological hazards; (ii) the discovery or rediscovery of local or traditional foods including wild and edible plants; (iii) increasing the participation of women in local agriculture through, for example, kitchen gardens, local kitchens, and cottage industries; (iv) improving the marketing and distribution of fresh foods in general and more specifically improving linkages between local agriculture and tourism. Together these strategies can help to increase food production and hence food self-sufficiency, reduce the need for food imports, improve availability of nutritious foods, increase value added for farmers, thereby improving economic viability and rural livelihoods, and improve the sustainability of local small-scale food systems.

Policies and strategies to improve food security the Caribbean should be informed by high-quality research. Individual Caribbean governments and CARICOM should therefore draw on the expertise of the local research community and involve the University of the West Indies as a partner. In the past 25 years or so, there has been significant research in the region on the topic of food and agriculture including research about renewable and sustainable agriculture, food security, traditional knowledge and agriculture, hazards, and local agriculture, among others.

Current research provides an ideal starting point for dissecting the issue of regional food security, but there is a need for ongoing research in a number of vital areas including ongoing analyses of the extent of food insecurity and hunger in the Caribbean; enhancement of the role, place, and fortunes of women in agriculture to serve the goals of improved food security and nutrition; ecological and economic sustainability of small-scale farming systems in the region; the role and potential of local knowledge and epistemologies in enhancing food security and nutrition in the region; identifying the main obstacles faced by farmers in increasing the production of affordable, nutritious food; and the marketing and distribution of locally grown fresh foods with a focus on strategies for strengthening linkages between the local agriculture and tourism sectors.

Bibliography

Abdool, A. and Carey, B. Making all-inclusives more inclusive: a research project on the economic impact of the all-inclusive sector in Tobago. Available at http://www.thetravelfoundation.org.uk/documents/All-inclusivesfinalreport-June04.doc.

Adger, N. Are there social limits to adaptation to climate change? *Climatic Change* 93, no. 3–4 (2009):335–354.

Adger, N., Arnell, N. and Tompkins, E. Successful adaptation to climate change across scales. *Global Environmental Change*, 15(2005):77–86.

Adger, N., Huq, S., Brown, K., Conway, D. and Hulme, M. Adaptation to climate change in the developing world. *Progress in Development Studies* 3(2003):179–195.

Agrawal, A. Dismantling the divide between indigenous and scientific knowledge. *Development Change* 26 (1995):413–439.

———. Why 'indigenous' knowledge? *Journal of the Royal Society of New Zealand* 39 (2009):157–158.

Agrawal, P. and Mall, R. Climate change and rice yields in diverse agro-environments of India. II. Effect of uncertainties in scenarios and crop models on impact assessment. *Climatic Change* 52(3)(2002):331–343.

Ahmed, B. (2001). The impact of globalization on the Caribbean sugar and banana industry. The Society for Caribbean Studies Annual Conference Paper 2.

Ahmed, B. and Afroz, S. *The Political Economy of Food and Agriculture in the Caribbean*. Kingston, Jamaica: Ian Randle, 1996.

Ambrose-Oji, B. Urban food systems and African indigenous vegetables: Defining the spaces and places for African indigenous vegetables in urban and peri-urban agriculture. In Shackleton, C.M., M.W. Pasquini and A. W. Drescher eds. *African Indigenous Vegetables in Urban Agriculture*. London: Earthscan, 2009, 1–33.

Anthes, R., Corell, R., Holland, G., Hurrell, J., McCracken, M. and Trenberth, K. Hurricanes and global warming – Potential linkages and consequences. *Bulletin of American Meteorological Society* 87, no.1 (2006):623–628.

Augier, R., Gordon, S.C., Hall, D.G. and Reckford, M. *The Making of the West Indies*. London: Longman, 1960.

Axline, W.A. Agricultural co-operation in CARICOM. In Payne, A. and P. Sutton, eds. *Dependency under Challenge: The Political Economy of the Commonwealth Caribbean*. Manchester: Manchester University Press, 1984, 152–172.

Bailey, S.W. Farmers perception of land degradation in the Pindars river valley, Clarendon. *Caribbean Geography* 13, 2 (2003):130–44.

Barbados Advocate. FAO examines high Caribbean food import bill. *Barbados Advocate* January 4, 2012. Retrieved February 19, 2013 from www.barbadosadvocate.com/newsitem.asp?more=local&NewsID=21896.

Barker, D. Dualism and disasters on a typical island: constraints on agricultural development in Jamaica. *Tijdschrift voor Economische en Sociale Geografie* 84, no.5 (1993):332–340.

———. Yam farmers on the edge of Cockpit Country: Aspects of resource use and sustainability. In McGregor, D., D. Barker and S. Evans eds. *Resource Sustainability and Caribbean Development*. Kingston: The University of the West Indies Press, 1998, 357–372.

———. A periphery in genesis and exodus: reflections on rural-urban relations in Jamaica. In Potter, R.B. and T. Unwin eds. *The Geography of Urban-Rural Interaction in Developing Countries*. London: Routledge, 1989, 294–322.

———. The Hazards of dryland farming on the plains of southern St Elizabeth, Jamaica. *Carib Xplorer-Life and Science of the Caribbean*, 1, no.3 (2010):22–26.

Barker, D. and Beckford, C.L. Plastic yam and plastic yam stick-Perspectives on indigenous technical knowledge among Jamaican farmers. *Tijdschrift voor Economische en Sociale Geographie*, 97, no.5 (2006):535–546.

———. Agricultural Intensification in Jamaican small-scale farming systems: vulnerability, sustainability and global change. *Caribbean Geography* 15 (2008):160–170.

———. The Yam Stick Problem in Jamaica: the role of farmers' knowledge the search for sustainable solutions. Paper presented at the Institute of British Geographers Workshop on Sustainable Development, 2003.

———. Yam production and the yam stick trade in Jamaica: Integrated problems for resource management. In McGregor, D. and D. Barker eds. *Resources, Planning, and Environmental Management in a Changing Caribbean*. Kingston: The University of the West Indies Press, 2003, 57–73.

Barker, D. and Specnce, B. Afro-Caribbean agriculture: A Jamaican Maroon community in transition. *The Geographical Journal* 154, no.2 (1988):198–208.

Barker, D., Collymore, J. and Spence, B.A. Learning from peasant farmers: Some Caribbean examples. Presented at the National Conference for Geographical Education, Ocho Rios, Jamaica, October 1983.

Barker, D., Dodman, D. and McGregor, D.F.M. Caribbean Vulnerability and Global Change: Contemporary Perspectives. In McGregor, D.F.M., D. Dodman and D. Barker, eds. *Global Change and Caribbean Vulnerability: Environment, Economy and Society at risk*. Kingston: University of the West Indies Press, 2009, 3–21.

Barrow, C. Small farm food production and gender in Barbados. In J.H. Momsen ed. *Women and Change in the Caribbean: A Pan Caribbean Perspective*. Kingston: Ian Randle Publishers, 1993, 181–193.

Bebbington, A. Capitals and capabilities: A framework for analyzing peasant viability, rural livelihoods and poverty. *World Development* 27, no.12 (1999):2021–2044.

Bechard, M. Feeding the world: scaling up projects will tackle food security through an interdisciplinary approach. *UniWorld*. Association of Universities and Colleges in Canada, March 2010, 5–7.

Beckford, C. Decision-making and innovation among small-scale yam farmers in central Jamaica; a dynamic, pragmatic and adaptive process. *The Geographical Journal*, 168, no.3 (2002):248–259.

————. Issues in Caribbean food security: Building capacity in local food production systems. In Aladjadjiyan, A. ed. *Food Production: Approaches, Challenges and Tasks*. Rijeka, Croatia: InTech, 2012.

————. Yam cultivation, the yam stick trade and resource depletion in the yam growing region of central Jamaica: Integrated problems for planning and resource management. PhD Dissertation, Department of Geography, University of the West Indies, Mona, 2000.

Beckford, C. and Barker, D. The role and value of local knowledge in Jamaican agriculture: Adaptation and change in small-scale farming. *The Geographical Journal* 173, no.2 (2007):118–128.

Beckford, C., Barker, D. and Bailey, S. Adaptation, innovation and domestic food production in Jamaica: Some examples of survival strategies of small-scale farmers. *Singapore Journal of Tropical Geography* 28, no.1 (2007):273–286.

Beckford, C.L. and Bailey, S.W. Vulnerability, constraints and survival on small-scale food farms in St Elizabeth, Jamaica: Strengthening local food production systems. In McGregor, D.F.M., D. Dodman and D. Barker, eds. *Global Change and Caribbean Vulnerability: Environment, Economy and Society at Risk*. Kingston, Jamaica: The University of the West Indies Press, 2009, 218–236.

Beckford, C.L., Campbell, D. and Barker, D. Sustainable food production systems and food security: Economic and environmental imperatives in yam cultivation in Trelawny, Jamaica. *Sustainability* 3 (2011):541–561.

Beckford, G. *Persistent Poverty Underdevelopment in Plantation Economies of the Third World*. New York: Oxford University Press, 1972.

————. Caribbean rural economy. In Beckford, G. ed. *Caribbean Economy: Dependence and Backwardness*. Kingston, Jamaica: Institute of Social and economic Research, University of the West Indies, 1975, 77–91.

————. Toward an appropriate theoretical framework for agricultural development planning and policy. *Social and Economic Studies* 3, no. 17 (1968): 233–242.

Belisle, F.J. Tourism and food production in the Caribbean. *Annals of Tourism Research* 10, 4 (1983):497–513.

————. The significance and structure of hotel food supply in Jamaica. *Caribbean Geography* 1, no. 4 (1984):219–233.

Benito, C.A. Cultural action and rural development. Rural Development Project Working Paper #9. Berkley, CA: University of California Press.

Berkes, F. Indigenous ways of knowing and the study of environmental change. *Journal of the Royal Society of New Zealand* 39 (2009):151–156.

Berleant-Schiller, R. and Pulsipher, L.D. Subsistence cultivation in the Caribbean. *New West Indian Guide* 60, no. 1–2 (1986):1–40.

————.Family land and Caribbean society: Towards an ethnography of Afro-Caribbean peasantries. In Thomas-Hope, E.M. ed. *Perspectives on Caribbean Regional Identity*. Liverpool: Liverpool University Press, 1984, 57–83.

Bernal, R., Figuerora, M. and Witter, M. Caribbean economic thought: The critical tradition. *Social and Economic Studies* 3, no. 2 (1984):5–96.

Besson, J. Gender and developmentin the Jamaican Small-Scale Marketing System: From the 1660s to the Millennium and Beyond. In Barker, d and McGregor, D.M.F. eds. *Resources, Planning and Environmental Management in a Changing Caribbean.* Kingston:University of the West Indies Press, 2003, 11–35.

————. Consensus in the family land controversy: rejoinder to Michaeline A. Crichlow. *New West Indian Guide/Nieuwe West-Indische Gids* 3–4 (1995):299–304.

Best, L. *The Choice of Technology Appropriate to Caribbean Countries.* Montreal: Mcgill University Center for Developing-Area Studies, 1976.

Blaikie, P.M. The theory of the spatial diffusion of innovations: a spatial cul-de-sac. *Progress in Human Geography*, 2 (1978):268–95.

Blaut, J. A study of cultural determinants of soil erosion and conservation in the Blue Mountains of Jamaica. *Social and Economic Studies* 5, no.2 (1959):403–420.

Bliss, E. Adaptation to agricultural change among Garifuna women in Hopkins, Belize. *Caribbean Geography*, 3, no.2 (1992):143–159.

Bloem, M., Huq, N., Gorstein, J., Burger, S., Khan, T., Islam, N., Baker, S. and Davidson, F. Production of fruits and vegetables at the homestead is an important source of vitamin A among women in rural Bangladesh. *European Journal of Clinical Nutrition* 50, 3 (1996):62–67.

Bocage, C. 'Status of Implementation of food and Nutrition Security Program'. Country paper presented at Seminar on The Role of Communication Tools for Implementing Food and Nutrition Security Plans in the Caribbean. CARDI/CTA/CFNI, Belize, 2005.

Boothroyd, J. Healthy food for the picking: Kitchens and wild plants could be key to Lebanon's food security. *UniWorld* Association of Universities and Colleges in Canada, 2010, 14–15.

Boxhall, R.A. Storages losses. In Gobob, P.G. Farrell and J.E. Orchard eds. *Crop-Post-Harvest: Science and Technology Volume.* Oxford: Blackwell Sciences, Ltd, 2002, 143–169.

Briggs, J. The use of indigenous knowledge in development: problems and challenges. *Progress in Development Studies*, 5(2005):99–114.

Briggs, J. and Sharp, J. Indigenous knowledges and development: A post colonial caution. *Third World Quarterly*, 25, no. 4 (2004):661–676.

Brohman, J. New directions in tourism for Third World development. *Annals of Tourism Research*, 23 (1996):48–70.

Brokenshaw, D.W., Werner, D.M. and Werner, O. *Indigenous Knowledge Systems and Development.* Lanham, MA: University Press of America, 1980.

Brooks, N. and Adger, N. *Country Level Risk Measures of Climate-Related Natural Disasters and Implications for Adaptation to Climate Change.* Norwich: Zuckerman Institute for Connective Environmental Research, 2003.

Brown, I. Impact of climate change on Caribbean agriculture. *Jamaica Observer*, 2005.

Brown-Glaude, W. Higglers in Kingston: Women's Informal Work in Jamaica. Nashville: Vanderbilt University Press, 2011.

Brownrigg, L. *Home Gardening in International Development: What the Literature Shows*. Washington, DC: The League for International Food Education, 1985.

Bunch, R. Encouraging farmers' experiments. In Chambers, R., A. Pacey and L. Thrupp eds. *Farmer First: Farmer Innovation and Agricultural Research*. London: ITDG, 1989, 55–61.

Burton, I., Kates, R.W. and White, G.F. *The Environment as Hazard*. New York: Oxford University Press, 1978.

Buzby, J.C., Wells, H.F., Axtman, B. and Mickey, J. *Supermarket Loss Estimates for Fresh Fruit, Vegetables, Meat, Poultry and Seafood in Their Use in the ERS Loss Adjusted Food Availability Data*. Economic Information Bulletin No. 44. Washington, DC: US Department of Agriculture, Economic Research Service, 2009.

Campbell-Chin Sue, H., Fielding, W.J. and Pacheco, A.R. *Adoption of Minisett Technology in Jamaica*. Kingston: International Institute for Cooperation in Agriculture, 1995.

Campbell, D. and Beckford, C.L. Negotiating uncertainty: Jamaican small scale farmers' adaptation and coping strategies before and after hurricanes. A case study of Hurricane Dean. *Sustainability* 1 (2009):1366–1387.

Campbell, D., Barker, D. and McGregor, D.F.M. Dealing with drought: Small farmers and environmental hazards in southern St Elizabeth, Jamaica. *Applied Geography* (2010):1–16.

Campbell, D.R. Domestic food production and hazard vulnerability in Jamaica: Adaptation and change in southern Saint Elizabeth. Ph. D. dissertation, Department of Geography and Geology, University of the West Indies, Mona, 2011.

Campesino Economic Organizations (CIOEC) Towards Food Sovereignty: Constructing an Alternative to the World Trade Organization's Agreement on Agriculture (2003). Available at www.nffc.net/.../Foodsovereinty_anAlternative.pdf.

Campillo, F. and Kleysen. *Productoras de alimentos en 18 paises de America Latina y el Caribe: Sintesis Hemisferica*. San Hose, CR: International Institute for Coorporation in Agriculture and International Development Bank, 1996, 101.

CARICOM View (2011). Food Security in CARICOM. *CARICOM View, July 2011*.

CARICOM Secretariat. *Regional Food and Nutrition Security Policy*. CARICOM, 2010.

CARICOM Secretariat. Agricultural Development Profile-Jamaica. 2009. Available at http://www.caricom.org/jsp/community/donor_conference_agriculture/agri_profile_jamaica.jspCaribbean Food and Nutrition Institute (CFNI)/Food Agricultural Organization (FAO). *Vulnerability and Food and Nutrition Security in the Caribbean*.CFNI/FAO, 2007.

CFNI/USDA/CARICOM Secretariat. Policy Dialogue-Technical Meeting on Poverty Alleviation and Food Security Strategies in the Caribbean, 2003.

Caribbean Food Emporium.'Caribbean Food Security'. 2003. Available at http://www.caribbeanfoodemporium.co.uk/foodsecurity.htm.

Carney, D. *Sustainable Rural Livelihoods, Environment and Development: Putting Poor Rural People First*. Sussex: Institute of Development Studies (IDS), 1998.

Carswell, G. Environmental problems in the tropics; Challenging the orthodoxies. In Jones, S. and Grace Carswell eds. *Environment, Development and Rural Livelihoods*. London: Earthscan, 2004, 3–11.

Cashman, K. Indigenous knowledge and international agricultural research: Where do we go from here? In Warren, D.M., J. Slikkerveer and S.O. Titilola eds. *Indigenous Knowledge Systems: Implications for Agriculture and International Development*. Ames, IA: Technology and Social Change Program, Iowa State University, 1989, 10–20.

CEDERA. Climate change and disaster management. Paper presented at the Adaptation to Climate Change in the Caribbean Disaster Risk Management Brainstorming Workshop, Barbados, 2002.

Center for Universal Education at Brookings. A Global Compact on Learning: Taking action on Education in Developing Countries. Washington D.C.: CUE, 2011.

Central Intelligence Agency (CIA). CIA Factbook Population rankings, 2012Available at https://www.cia.gov/library/publication/the-world-factbook/rankorder/2119rank.html.

Chambers, R. The World Development Report: Concepts, contents and a Chapter 12. *Journal of International Development*, 13 (2001):299–306.

———. *Whose Reality Counts? Putting the First Last*. London: Intermediate Technology Publications, 1997.

———. *Rural Development: Putting the Last First*. Harlow, UK: Longman, 1983.

Chambers, R., Pacey, A. and Thrupp, L.A. *Farmer First: Farmer Innovation and Agricultural Research*. London: Intermediate Technology Publications, 1989.

Charveriat, C. Natural Disasters in Latin America and the Caribbean: An Overview of Risks [Electronic Version, 2000]. Inter-American Development Bank Working Paper 434. Retrieved August 3, 2009 from http://www.iadb.org/sda/doc/ENVNatDisastLACeline.pdf.

Chen, S. and Ravillion, M. *How Have the World's Poorest Fared Since the Early 1980s?* Washington, D.C.: World Bank Development Research Group, 2004. www.worldbank.org/research/povmonitor/MartinPapers/How_have_the_poorest_fared_since_the_early_1980s.pdf.

Chen, A. and Taylor, M. Investigating the link between early season Caribbean rainfall and the ELnino + 1 year. *International Journal of Climatology*, 22(2002):87–106.

Central Intelligence Agency (CIA). CIA The World Factbook, 2012. Available at https://www.cia.gov/library/publications/the-world-factbook/geos/bf.html.

Cleaver, F. Paradoxes of participation: Questioning participatory approaches to development. *Journal of International Development* 11 (1999):597–612.

Cofie, O., Bradford, A.A. and Drechsel, P. Recycling of urban organic waste for urban agriculture. In van Veenhuizen, R. ed. *Cities Farming for the Future: Urban Agriculture for Green and Productive Cities*. The Philippines: RUAF Foundation, IDRC and IIRR, 2006, 209–242.

Cofie, O., van Veenhuizen, R. and Drechsel, P. Contribution of urban and peri-urban agriculture to food security in sub-Saharan Africa. Paper presented at the Africa session of the 3rd WWF Symposium, Kyoto March 17, 2003.

Collier, P. The politics of hunger. *Foreign Affairs* 87, no. 6 (2008):67–79.

————. Peasant agriculture is incompatible with economic development: Africa needs GM crops. Available at http://www.africanagricultureblog.com/2009/10/peasant-agriculture-is-incompatible.html.

Collymore, J. Agricultural decisions of small farmers in St Vincent. MPhil thesis, department of Geography, University of the West Indies, Mona, 1984.

————. Integrating farmers into environmental monitoring programs. Paper presented at the Workshop on Environmental development, Vieux Fort, St Lucia.

————. Small farmers as resource managers: A note from St. Vincent. *Caribbean Geography* 2 (1986):92–99.

Conway, D. Tourism, environmental conservation and management and local agriculture in the Eastern Caribbean: Is there an appropriate, sustainable future for them? In Duval, D.T. ed. *Tourism in the Caribbean: Trends, Development and Prospects.* London and New York: Routledge, 2004, 187–204.

Cooper, P., Dimes, J., Rao, K., Shapiro, B., Shiferaw, B. and Twomlow, S. Coping better with current climatic variability in the rain-fed farming systems of sub-Saharan Africa: An essential first step in adapting to future climate change? *Agriculture, Ecosystems and Environment,* 126, no.1–2 (2008):24–35.

Crichlow, M. An alternative approach to family land tenure in the anglophone Caribbean: The case of St. Lucia. *New West Indian Guide,* 68 (1994):77–99.

Davis-Morrison, V. The sustainability of small-scale agricultural systems in the Millbank area of the Rio Grande Valley, Portland, Jamaica. In McGregor, D., D. Barker and S. Evans eds. *Resource Sustainability and Caribbean Development.* Kingston: The University of the West Indies Press, 1998, 296–316.

Davis-Morrison, V. and Barker, D. Resource management, environmental knowledge and decision-making in small farming systems in the Rio Grande Valley, Jamaica. *Caribbean Geography,* 8, no.2 (1997):96–106.

De Lucia, M. and Assennato, D. Agricultural engineering in development: Postharvest operations, management of foodgrains. *FAO Agricultural Services Bulletin No. 93.* Rome: FAO.

Deep Ford, J.R. Towards a new food policy strategy for the Caribbean: Linking food security, health and surveillance. Presented at the Policy Dialogue-Technical Meeting on Poverty Alleviation and Food Security Strategies on the Caribbean. Caribbean Food and Nutrition Institute and United States Development Agency: Kingston, Jamaica November 19–21, 2003.

Deep Ford, J.R., Crescenzo dell'Aquila, and Piero Conforti, eds. Agricultural trade policy and food security in the Caribbean: Structural issues, multilateral negotiations and competitiveness. Rome: Food and Agricultural Organization, 2007.

Deep Ford J.R. and Khaira, H. Caribbean countries as small and vulnerable economies in the WTO. In Deep Ford, J.R., Crescenzodell'Aquila, and PieroConforti, eds. *Agricultural trade policy and food security in the Caribbean: Structural issues, multilateral negotiations and competitiveness.* Rome: Food and Agricultural Organization, 2007, 41–71.

Deep Ford, J.R. and Rawlins, G. Trade policy, trade and food security in the Caribbean. In Deep Ford, J.R., Crescenzodell'Aquila, and PieroConforti, eds. *Agricultural trade policy and food security in the Caribbean: Structural issues, multilateral negotiations and competitiveness*. Rome: Food and Agricultural Organization, 2007, 7–39.

Demas, W. G. Situation and change. In Beckford, G. ed. *Caribbean Economy: Dependence and Backwardness*. Kingston, Jamaica: Institute of Social and economic Research, University of the West Indies, 1975, 61–76.

Demas, W. The prospects for developing agriculture in small Commonwealth Caribbean territories: The role of small-scale farmers. In Proceedings of the Fifth West Indian Agricultural Economics Conference. Port of Spain: University of the West Indies, St Augustine.

Dixon, D. Wetland sustainability and the evolution of indigenous knowledge in Ethiopia. *Geographical Journal*, 171 (2005):306–323.

Dodman, D. and Rhiney, K. We nyammin: Food authenticity and the tourist experience in Negril, Jamaica. In Daye, M., D. Chambers and S. Roberts eds. *New Perspectives in Caribbean Tourism*. New York: Routledge, 2008, 115–132.

Durant-Gonzalez, V. Higglering: Rural women and the internal marketing system in Jamaica. In Gomes, I. P. ed. *Rural Development in the Caribbean*. Kingston, Jamaica: Heinemann Educational Books (Caribbean) Ltd, 1985, 103–122.

———. The occupation of higglering. *Jamaica Journal* 16 no3, (1983):2–12

EC-FAO Food Security Program: An introduction to the Basic Concepts of Food Security. Food Security Information for Action Practical Guidelines, 2008

ECLAC. Wither the Agricultural Sector in the Caribbean? Economic Commission for Latin America and the Caribbean.[Electronic Version 2004]. Retrieved June 14, 2008 from: http://www.eclac.org/portofspain/noticias/bolnoticias/9/26869/issue19.pdf.

Edwards, D.T. *Report on an Economic Study of Small Farming in Jamaica*. Institute of Social and Economic Research (ISER), University College of the West Indies, Mona, 1961.

Emanuel, K. The dependence of hurricane intensity on climate. *Nature*, 326(1987):483–485.

Emeagwali, G. African Indigenous Knowledge Systems (AIK): Implications for the curriculum. In Falola, T. ed. *Ghana in Africa and the World: Essays in Honour of Adu Boahen*. New Jersey: African World Press, 2003.

Eriksen, S., Brown, K. and Kelly, M. The dynamics of vulnerability: locating coping strategies in Kenya and Tanzania. *The Geographical Journal* 171, no.4 (2005):287–305.

Evans, P. *Agro-Forestry Development in the Yam Growing Region of Central Jamaica*. Rome: FAO, 1994.

Eyzagurrie, P. B. Global recognition of indigenous knowledge: Is this the latest phase in globalization? *Indigenous Knowledge and Development Monitor* 9 (2001). Available at http://www.nuffic.nl/ciran/ikdm/9–2/column.html.

FAO. *The State of Food and Agriculture: Women in Agriculture.* FAO: Rome. Available at http://www.fao.org/publications/sofa/en/

———. Statistical Yearbook, 2012: World food and agriculture. Rome: FAO, 2012

———. *State of Food and Agriculture. Women in Agriculture: Closing the Gender Gap for Development.* FAO: Rome, 2011

———. *The Female Face of Farming: Farming First.* FAO: Rome, 2012. Available at http://www.guardian.co.uk/news/datablog/1022/mar/02/female-face-farming-fao#zoomed-picture at Work

———. *Potentials for Agriculture and Rural Development in Latin America and the Caribbean.* Rome: United Nations, 1988.

———. *The State of Food Insecurity in the World.* Rome: United Nations, 1999.

———. *Report of the Panel of Eminent Experts on Ethics in Food and Agriculture.* Rome: United Nations, 2003.

———. *Improving Nutrition through Home Gardening: A Training Package for Preparing Field Workers in Southeast Asia, Food and Nutrition Division.* Rome: FAO, 1995.

———. *The Special Program for Food Security. Urban and Peri-urban Agriculture: A Briefing Guide for the Successful Implementation of Urban and Peri-urban Agriculture in Developing Countries of Transition.* Rome: FAO, 2001.

———. *The State of Food and Agriculture in the World, 2007.* Rome: FAO, 2008.

———. *The State of food and Agriculture, Agricultural Trade and Poverty.* Rome: FAO, 2005b.

———. *The State of Food Insecurity in the World: How does international price volatility affect domestic economies and food security?* Rome: FAO, 2011.

———. *Women-Key to Food Security.* Rome: FAO, 2010, Available at http://www.fao.org/docrep/014/am719e/am719e00.pdf

———. *Crop Prospects and Food Situation.* Rome: FAO, 2008.

FAO-COAG. The COAG-Paper. Report of the COAG Secretariat to the COAG. Rome: FAO, 1999.

Figueroa, M. The Plantation School and Lewis: Contradictions, continuities and continued Caribbean relevance. *Social Economic Studies*, 45, no. 4 (1996): 23–49.

Finerman, R. and Sackett, R. Using homegardens to decipher health and healing in the Andes. *Medical Anthropology Quarterly* 17, 4 (2003):459–82.

Flora, C. *Interactions between Agroecosystems and Rural Communities. Advances in Agroecology Series.* Boca Raton: CRC Press, 2001.

Floyd, B.N. No easy harvest: Policies and priorities for agriculture in Jamaica. *Journal of Geography* 82, no.5 (1983):212–221.

Forsyth, T. Science, myth and knowledge: Testing Himalayan environmental degradation in Thailand. *Geoforum* 27 (1996):375–92.

Found, W.C. *A Theoretical Approach to Rural Land-use Patterns.* London: Edward Arnold, 1971.

Friends of the Earth. The Economic Benefits of Farmers' Markets, 2011. Available at http://www.foe.co.uk/resource/breifings/farmers_markets.pdf.

Gamble, D. Caribbean vulnerability: An appropriate climatic framework. In McGregor, D., D. Dodman and D. Barker eds. *Global Change and Caribbean Vulnerability: Environment, Economy and Society at Risk.* Kingston: University of the West Indies Press, 2009.

Gamble, D., Campbell, D., Allen, T., Barker, D., Curtis, S., McGregor, D. and Popke, J. Climate change, drought, and Jamaican agriculture: local knowledge and the climate record. *Annals of the Association of American Geographers* 100, no. 4 (2010):880–893.

Gates Foundation. Women in Agriculture 2012. Available at www.gatesfoundation. org/infographics/Pages/women-agriculture-info.aspx.

Gbetibouo, G. Understanding Farmers' Perceptions and Adaptations to Climate Change and Variability' The Case of the Limpopo Basin, South Africa [Electronic Version 2009]. IFPRI Discussion Paper 00849. Retrieved January 17, 2010 from http://www.ifpri.org/pubs/dp/IFPRIDP00849.pdf.

Ghatak, S. and Ingersent, K. *Agriculture and Economic Development.* Baltimore: Johns Hopkins University Press, 1984.

Girvan, N. and Jefferson, O. *Readings in the Political Economy of the Caribbean.* Kingston, Jamaica: Tapia House and the New World Group, 1971.

Gliessman, S.R. *Agroecology: Ecological Processes in Sustainable Agriculture.* Boca Raton: Lewis/CRC Press, 1998.

Agroecosystem Sustainability: Toward Practical strategies. Advances in Agroecology Series. Boca Raton: CRC Press, 2001.

Global Conference on Women in Agriculture (GCWA). Press Release. Africa: Neglecting prominent role of women in Agriculture hindering ability to resolve threats to food security. Indian Council of Agricultural Research (ICAR) and Asia-Pacific Association of Agricultural Research Institutions (APAARI) 2012. Available at http://allafrica.com/stories/201203121548.html.

―――. Synthesis Report. Indian Council of Agricultural Research (ICAR) and Asia-Pacific Association of Agricultural Research Institutions (APAARI) 2012.

Goldenberg, S., Landsea, W., Mestas-Nu–ez, A. and Gray, W. The recent increase in Atlantic hurricane activity: causes and implications. *Science* 293 (2001): 474–479.

Gooding, E.B. ed. *Pest and Pesticide Management in the Caribbean.* Bridgetown: Consortium for International Crop Protection, 1980.

Gonzalez, N.L. *Black Carib Household Structure: A Study of Migration and Modernization.* Seattle: University of Washington Press, 1969.

Sojourners of the Caribbean: Ethnogenisis and Ethnohistory of the Garifuna. Urbana: University of Illinois Press, 1988.

Greenwood, R. and Hamber, S. *Emancipation to Emigration.* London: Macmillan Education Ltd, 1980.

Grothman, T. and Pratt, A. Adaptive capacity and human cognition: The process of individual adaptation to climate change. *Global Environmental ChangePart A,* 15, 3 (2005):199–213.

Grossman, L.S. Pesticides, people and environment in St Vincent. *Caribbean Geography* 3, no.3 (1992):175–186.

Gyampoh, B., Amisah, S., Idinoba, M. and Nkem, J. Using traditional knowledge to cope with climate change in rural Ghana. *Unasylva*, 60, no. 231/232 (2008): 70–74.

Hagerstrand, T. *Innovation Diffusion as a Spatial Process*. Chicago: University of Chicago Press, 1967.

Haggblade, S. The rural non-farm economy: Pathway out of poverty or pathway in? The Future of Small Farms. IFPRI, ODI. Wye, Kent: Imperial College, 2005.

Hahn, M., Riederer, A. and Foster, S. The Livelihood Vulnerability Index: A pragmatic approach to assessing risk from climate variability and change – a case study in Mozambique. *Global Environmental Change*, 19(2009):74–88.

Halweil, B. Farmland defense: How the food system can ward off future threats. In New perspectives on food security Glynwood Center Conference Proceedings, Cold Spring, New York: Glynwood Center, 2005, 25–30.

Halweil, B. and Thomas, P. *Home Grown: The Case for Local Food in a Global Market*. Washington, DC: World Watch Institute, 2012.

Harris, S.R. *Production Is Only Half the Battle: A Training Manual in Fresh Produce Marketing for the Eastern Caribbean*. FAO: Rome, 1988.

Hawkins, I. *The Changing Face of the Caribbean Bridgetown, Barbados*. Bridgetown: CEDAR Press, 1979.

Hayle, C. Tourism in Jamaica: The impact of the Past and the Future. In Jayawardena, C. ed. *Caribbean Tourism: Visions, Missions and Challenges*. Kingston: Ian Randle Publishers, 2005, 119–139.

Helen Keller International/Asia-Pacific. Integration of Animal Husbandry into Home Gardening Programmes to Increase Vitamin A in take from Foods: Bangladesh, Cambodia and Nepal. 2003. Retrieved from http://www.hkiasiapacific.org/_downloads/APOpercent20Specialpercent20issueJanpercent202003.pdf.

———. Homestead Food Production – A Strategy to Combat Malnutrition and Poverty. Helen Keller International: Jakarta, Indonesia, 2001.

Hills, T.L. The Caribbean Food forest: Ecological artistry or random chaos? In Brierley, J.S. and H. Rubenstein eds. *Small Farming and Peasant Resources in the Caribbean*, Manitoba Geographical Studies 10, University of Manitoba, 1988, 1–28.

———. A re-assessment of the traditional. *Caribbean Geography* 1 (1983):24–35.

Hills, T. and Iton, D. A re-assessment of the traditional in Caribbean small-scale agriculture. *Caribbean Geography* 1 (1983):24–35.

Hodges, R.J., Buzby, C.J. and Bennett, B. Postharvest losses and waste in less developed countries: opportunities to improve resource use. *Journal of Agricultural Science*, 149 (2010):1–9.

Holt-Gimenez, E. Movimiento Campesino a Campesino: Linking sustainable agriculture and social change. *Food first Backgrounder*, 12, no. 1 (2006).

Hoogerbrugge, I. and Fresco, L.O. Small home garden plots and sustainable livelihoods for the poor. FAO: Rome (1993). FAO Corporate Document Repository. Retrieved from http://www.fao.org/docrep/007/j2545e/j2545e01.htm #TopofPage.

Horowitz, M. *Morne-Paysan, Peasant Village in Martinique*. New York: Holt Rinehart and Winston, 1967.

Howes, M. The uses of indigenous technical knowledge in development. *IDS Bulletin* 10 (1979):12–23.

Howes, M. and Chambers, R. Indigenous technical knowledge: Analysis, implications and issues. *IDS Bulletin*, 10 (1979):5–11.

Hudson, N.W. Research and Training requirements in developing countries. In Grenland, D.J. and R. Lal eds. *Soil Conservation and Management in the Humid Tropics*. Chichester: Wiley, 1982, 11–16.

Huntingford, C., Lambert, F., Gash, J., Taylor, C. and Challinor, A. Aspect of climate change prediction relevant to crop productivity. *Philosophical Transactions of the Royal Society B*, 360 (2005):1999–2009.

Ilbery, B.W. Point score analysis: A methodological framework for analyzing the decision-making process in agriculture. *Tijdschrift voor Economische en Sociale Geographie*, 68 (1977):66–71.

———. *Agricultural Geography: A Social and Economic Analysis*. New York: Oxford University Press, 1985.

Igbozurike, M. An empirical contribution to the issue of farm size in land parcellization. *Journal of Geography*, 75, no.5 (1976):266–275.

Ingham, A., Ma, J. and Ulph, A. *How do the Costs of Adaptation Affect Optimal Mitigation When There is Uncertainty, Irreversibility and Learning?* London: Tyndall Centre for Climate Change Research, 2005.

Innis, D.Q. *Intercropping and the Scientific Basis of Traditional Agriculture*. London: ITDG, 1997.

———. The efficiency of Jamaican peasant land use. *Canadian Geographer*, 5 (1961):19–23.

———. The efficiency of tropical small farm agricultural practices. In Adams, W. p. and F.M. Helleiner eds. *International Geography*. Toronto: University of Toronto Press, 1972.

Institute for Agriculture and Trade Policy Towards Food Sovereignty: Constructing an alternative to the World Trade Organization's agreement on agriculture, farmers, food and trade. International Workshop on the Review of the AoA, 2003. Geneva. Retrieved from http://www.tradeobservatory.org/library.cfm?RefID=25961.

International Monetary Fund (IMF). World Economic Outlook, 2011. New York: IMF.

IPCC. *Fourth Assessment Report: Climate Change 2007-Synthesis Report*. Cambridge: Cambridge University Press, 2007.

———. Climate Change: The Scientific Basis Contribution of Working group 1 to the Third assessment Report of the Intergovernmental Panel on Climate Change; Houghton, J., Y. Ding, D. Griggs, M. Noguer, M. van der Linden, P. Dai, X. K. Maskell and C. Johnson eds. Cambridge: Cambridge University Press, 2001.

Ishemo, A., Semple, H. and Thomas-Hope, E. Population mobility and the survival of small farming in the Rio Grande Valley, Jamaica. *The Geographical Journal*, 172, no. 4 (2006):318–330.

Jackson, D.L. and Jackson, L.L. *The Farm as a Natural Habitat*. Washington DC: Island Press, 2002.

Jacobi, P., Drescher, A.W. and Amend, J. *Urban Agriculture–Justification and Planning Guidelines*. City Farmer, Canada's Office of Urban Agriculture, 2000.

Jama, B. *Learning from the farmer. Farmers and Agricultural Research: Complementary Methods*. IDS Workshop, University of Sussex, July, 1987.

Jayawardena, C. and Ramjee-Singh, D. Performance of tourism analysis: A Caribbean perspective. *National Journal of Contemporary Hospitality Management*, 5, no.3 (2003):176–79.

Jenkins, J. and Scanlan, S. Food Security in Less Developed Countries, 1970–1990. *American Sociological Review*, 66, no.5 (2001):718–744.

Jones, G.E. The diffusion of agricultural innovations. *Journal of Agricultural Economics*, 15 (1963):49–59.

———. *Innovation and Farmers' Decision-Making*. Agriculture Open University, Milton Keyes: Open University Press, 1975.

Jones, P., Horton, E., Folland, C., Hulme, M., Parker, D. and Basnett, T. The use of indices to identify changes in climatic extremes. *Climatic Change*, 42 (1999):131–149.

Kalland, A. Indigenous Knowledge: Prospects and limitations. In Ellen, R., P. Parkes and A. Bicker eds. *Indigenous Environmental Knowledge and its Transformations*. Amsterdam: Hardwood Academic Publishers, 2000, 1–33.

Kapoor, I. The devil is in the theory: A critical assessment of Robert Chambers' work on participatory development. *Third World Quarterly* 23 (2000): 101–117.

Kasperson, R. and Kasperson, J. *Climate Variability, Vulnerability and Social Justice*. Stockholm: Stockholm Environment Institute, 2001a.

Kent, G. Food and trade rights. *UN Chronicle*, Issue 3 (2001). Retrieved from http://www.un.org/Pubs/chronicle/2002/issue1/010p7.html.

Khalid, F. and Babb, R. Hazard and risk assessment from Hurricane Ivan (2004) in Grenada using Geographical Information Systems and remote sensing. *Journal of Maps Student Edition* (2008):4–10. ISSN 1755-2958.

Khor, M. The impact of globalization and liberalisation on agriculture and small farmers in developing countries: The experience of Ghana. 2006 (Publication). Retrieved September 28, 2008, from Third World Network: http://www.twnside.org.sg/title2/par/Ghana_study_for_IFAD_project_FULL_PAPER_rev1apr06.doc.

Kimber, C.T. Spatial patterning in the dooryard gardens of Puerto Rico. *Geographical Review*, 63, no.1 (1973):19–23.

Kossin, J., Knapp, K., Vimont, D., Murnane, R. and Harper, B. (2007). A globally consistent reanalysis of hurricane variability and trends [Electronic Version 2007]. *Geophysical Research Letters*, *34*, L4815. Retrieved June 2, 2007 from http://www.aos.wisc.edu/~dvimont/Papers/Kossin_2006GL028836.pdf.

Kumar, B.M. and Nair, P.K.R. The enigma of tropical homegardens. *Agroforestry Systems* 61 (2004):135–152.

Kumar, S.K. Role of the household economy in child nutrition at low incomes: A Case Study in Kerala. *Occasional Paper No. 95*. Department of Agricultural Economics, Cornell University. Ithaca, New York, 1978.

Lado, C. Sustainable environmental resource utilization: A case study of farmers' ethnobotanical knowledge and rural change in Bungoma district, Kenya. *Applied Geography* 24 (2004):282–302.

Lambert, R. Monitoring local food security and coping strategies: Lessons from information collection and analysis in Mopti, Mali. *Disasters* 18, no.4 (1994):333–343.

Lang, T. How new is the world food crisis? Thoughts on the long-term dynamics of food democracy, food control and food policy in the 21st century, 2009a. Paperpresented to the Visible Warnings: The world food Crisis in Perspective Conference April 3–4, 2009, Cornell University, Ithaca, NY.

LeFranc, E. Social structure, land use, and food availability in the Caribbean, 2006. Available at http://www.unu.edu/unupress/food/8F034E02.htm.

Lewis, W. A. Economic development with unlimited supplies of labor. The *Manchester School of Economic and Social Studies* 22 (1954):139–191.

———. The shifting fortunes of agriculture (1958). In Gersovitz, M. ed. *Selected Economic Writings of W. Arthur Lewis*. New York: New York University Press, 1983, 27–34.

Lin Nan. *The Study of Human Communication*. Indianapolis: Bobbs-Merrill, 1973.

Liwenga, E. Adaptive livelihood strategies for coping with water scarcities in the drylands of central Tanzania. *Physics and Chemistry of the Earth* 33 (2008):775–779.

Luers, A., Lobell., D., Sklar, L., Addams, L. and Matson, P. A method for quantifying vulnerability, applied to the agricultural system of the Yaqui Valley, Mexico. *Global Environmental Change* 13 (2003):255–267.

Lugo, A. Effects and outcomes of Caribbean hurricanes in a climate change scenario. *Science of the Total Environment* 262 (2000):243–251.

Mandle, J.R. *Patterns of Caribbean Development: An Interpretative essay on Economic Change*. New York: Gordon Breach, 1982.

Marin, A. Riders under storms: Contributions of nomadic herders' observations to analyzing climate change in Mongolia. *Global Environmental Change* 20 (2010):162–176.

Marsh, R. Household Gardening and food Security. Presented at the International Conference of Agricultural Economists on Food Security, Diversification and Resource Management: Refocusing the Role of Agriculture, 1997.

Marshall, W.K. Peasant development in the Qwest Indies since 1838. In Gomes, P.I. ed. *Rural Development in the Caribbean*. Kingston Jamaica: Heinemann Educational Books, 1985, 1–14.

McGregor, D. and Barker, D. Land degradation and hillside farming in the Fall River Basin, Jamaica *Applied Geography* 11 (1991):143–156.

McGregor, D., Barker, D. and Dodman, D. eds. *Global Change and Caribbean Vulnerability: Environment, Economy and Society at Risk*. Kingston: The University of the West Indies Press, 2009, 273–297.

McGregor, D.F.M., Barker, D. and Campbell, D. Environmental change and Caribbean Food security: Recent hazard impacts and domestic food production in Jamaica. In McGregor, D.F.M., D. Dodman and D. Barker eds. *Global Change and Caribbean Vulnerability: Environment, Economy and Society at Risk*. Kingston, Jamaica: The University of the West Indies Press, 2009, 197–217.

McLeman, R. and Smit, B. Migration as an adaptation to climate change. *Climatic Change* 76 (2006):31–53.

McMichael, P. A food regime genealogy. *Journal of Peasant Studies* 35 (2009b):139–169.

Meassick, M. The Importance of Agriculture. 2004. Available at http://dfid-agriculture-consultation.nri.org/maillists/growth-and.

Mehdi, B. Coping with Drought in the Caribbean. Advanced Integrated Water Resources Management course 2007. Retrived from: http://www.mcgill.ca/files/cariwin/Day10_Mehdi_1Drought_in_the_Caribbean.pdf

Meijerink, G. and Roza, P. The role of agriculture in development. Markets, chains, and sustainable development. Strategy and Policy Paper, no. 5. Wageningen: Stichting DLO, 2007. Available at: http://www.boci.wur.nl/UK/publications/

Meikle, P. Spatial-Temporal Trends in Root Crop Production and Marketing in Jamaica. Unpublished M.Phil. Thesis, University of the West Indies, Mona, Kingston Jamaica. 1992.

Meikle-Yaw, P. Globalization of agriculture: Effects on social and natural systems in rural communities in Jamaica. *Caribbean Geography* 14, no.1 (2005):40–54.

Mellor, J. *Globalization and the Traditional Role of Agriculture.* Rome: Food and Agriculture Organization, 2002.

Mercer, J., Kelman, I., Suchet-Pearson, S. and Lloyd, K. Integrating indigenous and scientific knowledge bases for disaster risk reduction in Papua New Guinea. *Geografiska Annaler*: Series B, 91 (2009):157–183.

Mercer, J., Kelman, I., Taranis, L. and Suchet-Pearson, S. Framework for integrating indigenous and scientific knowledge for disaster risk reduction. *Disasters*, 34 (2010):214–239.

Meyer, D. Caribbean tourism, local sourcing and enterprise development: review of literature. Pro-Poor Tourism. Working Paper #18, Center for Tourism and Cultural Change, Sheffield Hallam University and Travel Foundations, 2006.

Mimura, N., Nurse, L., McLean, R., Agard, J., Briguglio, L., Lefale, P., Payet, R., Sem, G. Small islands. In M. Parry, Canziani, O., Palutikof, J., van der Linden, P. and Hanson, C., ed. *Climate Change 2007: Impacts, Adaptation and Vulnerability. Contribution of Working Group II to the Fourth Assessment Report of the Intergovernmental Panel on Climate Change.* Cambridge, UK: Cambridge University Press, 2007, 687–716.

Mintz, S.W. *Caribbean Transformations.* Chicago: Aldine Publishing Co., 1974.

———. *Caribbean Transformations.* New York: Columbia University Press, 1989.

———. From plantations to peasantries in the Caribbean. In Mintz, S. ed. *Caribbean Contours.* Baltimore: Johns Hopkins University Press, 1985, 127–153.

Mintz, S.W. and Hall, D. The origins of the Jamaican internal marketing system. Yale University Publications in Anthropology, Yale University 1960.

Mintz, S.W. and Price, R. The Birth of African-American Culture: An Anthropological Approach. Boston: Beacon Press, 1992.

Mitchell, R. and Hanstad, T. *Small Homegarden Plots and Sustainable Livelihoods for the Poor.* FOA Livelihood Support Program (LSP). Rome: FAO, 2004.

Mohan, G. and Stokke, K. Participatory development and empowerment: The dangers of localism. *Third World Quarterly* 21 (2000):247–68.

Momsen, J.H. Caribbean tourism and agriculture: New linkages in the global era? In Klak, T. ed. *Globalization and Neo-liberalism: The Caribbean Context.* Lanham, MD: Rowman and Littlefield, 1998, 267–272.

———. The marginalization of women farmers: Gender and small-scale agriculture in Barbados 1838–1935. *Caribbean Geography* 6, no.1 (1995):52–60.

———. Linkages between tourism and agriculture: Problems for the smaller Caribbean economies. Seminar Paper #45, Department of geography, University of Newcastle upon Tyne, 1986.

———. Changing gender roles in Caribbean peasant agriculture. In Brierley, J.S. and H. Rubenstein eds. *Small Farming and Peasant Resources in the Caribbean.* Manitoba Geographical Studies 10. University of Manitoba. 1988, 83–100.

———. *Report on Vegetable Production and the Tourist Industry in St Lucia.* Calgary: University of Calgary, Department of Geography, 1972.

Momsen, J. The Geography of Landuse and Population in the Caribbean(with special reference to Barbados and the Windward Islands). Un pub. PhD dissertation, University of London. 1969.

Moscow, A. The contributions of urban agriculture to individual control and community enhancement. Master of Science thesis, International Agricultural Development. UC Davis, 1995.

Moser, C. The asset vulnerability framework: Reassessing urban poverty reduction strategies. *World Development* 26, no.1 (1998):1–19.

Mougeot, L. *Growing Better Cities: Urban Agriculture for Sustainable Development.* Ottawa: International Development Research Center, 2006.

———. Urban agriculture: Definition, presence, potentials and risks, and policy challenges. Presented at the International Workshop "Growing Cities, Growing Food." October 11–15, 1999.

Murphy, C. Cultivating Havana: Urban agriculture and food security in the Years of Crisis. Food First-Institute for Food and Development Policy, Development Report No. 12. Oakland, CA: Food First, 2006.

———. Sustainable Development and food Security: The Case of Cuba. Thesis, New College of California, San Francisco.

Mustafa, D. Reinforcing vulnerability? Disaster relief, recovery, and response to the 2001 flood in Rawalpindi, Pakistan. *Environmental Hazard* 5 (2003): 71–82.

Myers, J. Natural disasters threaten agriculture sector. Kingston: *Jamaica Gleaner,* June 6, 2006. Available at http://jamaica-gleaner.com/gleaner/20060612/business/business1.html.

Naess, L., Norland, I., Lafferty, W. and Aall, C. (2006). Data and processes linking vulnerability assessment to adaptation decision-making on climate change in Norway. *Global Environmental Change* 16 (2006):221–233.

Nair, P.K.R. Do tropical home gardens elude science, or is it the other way around? *Agroforestry Systems* 53 (2001):239–245.

———. *An Introduction to Agroforestry.* Dordrecht: Kulwer, 1993.

Netherlands Organization for International Cooperation in Higher Education/Indigenous Knowledge (NUFFIC). Best practices on indigenous knowledge. NUFFIC, 2003. Available at http://www.nuffic.nl/ciran/ikdm/9–2/column.html.

Ninez, V.K. Household gardens: Theoretical and policy considerations. *Agricultural Systems* 23 (1987):167–186. Ninez. V. K. The household garden as lifeboat. *CERIS*, 112 no. 19, 4 (1986):31–36.

———. Introduction to household gardens and small-scale food production. *Food and Nutrition Bulletin* 7, 3 (1985):1–5.

Nugent, R.A. *The Significance of Urban Agriculture*. City Farmer, Canada's Office of Urban Agriculture, 1997.

Hills, T.L. The Caribbean Food forest: Ecological artistry or random chaos? In Brierley, J.S. and H. Rubenstein eds. *Small Farming and Peasant Resources in the Caribbean*, Manitoba Geographical Studies 10, University of Manitoba, 1988, 1–28.

Nurse, L., Sem, G. Small Island states. In McCarthy, J., O. Canziani, N. Leary and D. Dokken eds. *Climate Change 2001: Impacts, Adaptation, and Vulnerability*. Cambridge UK: Cambridge University Press, 2001, 843–875.

Nurse, O.J. Small-scale farming in Barbados. Unpublished proceedings of the 5th West Indian Agricultural Economic Conference.

O'Loughlin, J. Diffusion. *Annals of the Association of American Geographers* 88, no.4 (1981).

Ohga, K. *World Food Security and Agricultural Trade*. Paris: Organisation for Economic Co-operation and Development (OECD), 1998.

Okoli, O. I., Igbokwe, M.C., Ene, L. S.O and Nwokodye, J. U. Rapid multiplication of yam by minisett technique. *National Root Crop Research Institute Research Bulletin* 2, Nigeria, 1992.

Parry, M. *Climate Change and World Agriculture*. London: Earthscan Publications Ltd, 1990.

Parry, M., Rosenzweig, C., Iglesiasc, A., Livermored, M. and Fischer, G. Effects of climate change on global food production under SRES emissions and socio-economic scenarios. *Global Environmental Change* 14 (2004):53–67.

Pasquini, M. and Young, E.M. Preface. In Shackleton, C.M., M.W. Pasquini and A.W. Drescher eds. *African Indigenous Vegetables in Urban Agriculture*. London: Earthscan, 2009, XXI–XXVI.

Pelling, M. and High, C. Understanding adaptation: what can social capital offer assessments of adaptive capacity? *Global Environmental Change* 15 (2005):308–319.

Pelling, M. and Uitto, J. Small island developing states: natural disaster vulnerability and global change. *Environmental Hazard* 3 (2001):49–62.

Pennicook, P. The all-inclusive concept: Improving benefits to the Jamaican economy. In Hall, K. and R. Holding eds. *Tourism: The Driver of Change in the Jamaican Economy?* Kingston: Ian Randle, 2006, 31–38.

Peterson, T., Taylor, M., Demeritte, R., Duncombe, D., Burton, S., Thompson, F., Porter, A., Mercedes, M., Villegas, E., Fils, R., Tank, A., Martis, A., Warner, R., Joyette, A., Mills, W., Alexander, L. and Gleason, B. Recent changes in climate extremes in the Caribbean region. *Journal of Geophysical Research*, 107, no.D21 (2002):4601.

Pielke, R., Rubiera, J., Landsea, C., Fernandez, M., Klein, R. Hurricane vulnerability in Latin America and the Caribbean: normalized damage and loss potential. *Natural Hazards Review* 4 (2003):101–114.

Pinderhuges, R., Murphy, C. and Gonzalez, M. *Urban agriculture in Havana, Cuba.* Urban Studies Program, San Francisco State University, 2000.

Poncelet, J. Disaster management in the Caribbean. *Disasters* 21, no.3 (1997):267–279.

Portilla, M. and Zuniga, H. The growing strength of rural women micro-entrepreneurs in Latin America and the Caribbean. COMUNIICA First edition, Second stage, January–April, 2007, 5–12.

Potter, R., Barker, D., Conway, D. and Klak, T. *The Contemporary Caribbean.* Boston, MA: Prentice Hall, 2004.

Pretty, J., Guijt, I., Schoones, I. and Thompson, J. Regenerating agriculture: the agroecology of low-internal input and community-based development. In Kirby, J., P. O'Keefe and L. Timberlake eds. *Sustainable Development.* London: Earthscan, 1999, 125–145.

Pulsipher, L.M. Geographical methods for assessing locations, techniques, crops and the idealized role of the Caribbean slave gardens. Presented at the Society for American Archeology Meeting, Atlanta, Georgia, 1989.

———. Enlish, Irish and African influence on the landscape of Seventeen Century Montserrat in the West Indies. Proceedings, Middle States Division, American Association of Geographers 13 (1979).

Putnam, R., Leonardi, R. and Nanetti, R. *Making Democracy Work: Civic Traditions in Modern Italy.* Princeton, NJ: Princeton University Press, 1993.

Rajasekaran, B. A Framework for Incorporating Indigenous Knowledge Systems into Agricultural Research and Extension Organizations for Sustainable Agricultural Development in India. Ph. D. dissertation. Center for Indigenous Knowledge for Agriculture and rural development, Iowa State University, 1992.

———. Indigenous technical practices in a rice-based farming system. Draft report prepared for the Center for Indigenous Knowledge for Agriculture and Rural Development, Iowa State University, Available at http://www.ciensin.columbia.edu/docs/004–195/004.html.

Ramjee-Singh, D. Import content of tourism: Explaining differences among Island States. *Tourism Analysis* 1, no. 1 (2006):33–44.

Reed, M. S., Dougill, A.J. and Taylor, M.J. Integrating local and scientific knowledge for adaptation to land degradation: Kalahari rangeland management options. *Land Degradation and Development* 18 (2007):249–268.

Rhiney, K. Globalization, tourism and the Jamaican Food Supply Network. In McGregor, D.F.M., D. Dodman and D. Barker eds. *Global Change and Caribbean Vulnerability: Environment, Economy and Society at Risk.* Kingston, Jamaica: The University of the West Indies Press, 2009, 237–258.

———. Agri-tourism linkages in Jamaica: Prospects and challenges. *Carib Xplorer* 1, no. 3 (2010–11):18–21.

———. Towards a new model for improved tourism-agriculture linkages? The case of two farmer cooperatives in Jamaica. *Caribbean Geography* 15, no.2 (2008):142–59.

Rhoades, R. The role of farmers in the creation and continuing development of agricultural technology and systems. In Chambers, R., A. Pacey and L. Thrupp eds. *Farmer First: Farmer Innovation and Agricultural Research*. London: ITDG, 1989.

Richards, P. *Indigenous Agricultural Revolution: Ecology and Food Production in West Africa*. London:Huchinson, 1985.

———. *The Thirsty Caribbean*. TierramŽrica Environment and Development 2010. Retrieved from: http://www.tierramerica.info/nota.php?lang=eng&idnews=3330&olt=464.

———. *Indigenous Agricultural Revolution*. London: Hutchinson, 1985.

Robinson, J.M. and Hartenfeld, J.A. *The Farmers' Market Book: Growing food, Cultivating Community*. Bloomington: Indiana University Press, 2007.

Rogers, E.M. Diffusion Innovation. New York: Free Press, 1995.

Ryder, R. Point score analysis of agricultural decision-making in the Dominican Republic. *Caribbean Geography* 4, (1993):2–16.

Saarinen, T.F. Perception of the environment. Commission on College Geography Resource Paper 5, Washington DC, Association of American Geographers, 1969.

Sarewitz, D., Pielke, R. and Kuykhah, M. (2003). Vulnerability and risks: Some thoughts from a political and policy perspective. *Risk Analysis* 23 (2003): 805–810.

Satchell, V.M. *From Plots to Plantations: Land Transaction in Jamaica 1866–1900*. Kingston: ISER Publications, University of the West Indies, 1990.

Sauer, C.O. *The Early Spanish Main*. Berkeley: University of California Press, 1966.

Satyavthi, C.T., Bhardwaj, C. and Brahmanand, P.S. Role of farm women in agriculture. *Gender, Technology and Development* 14, no.3 (2010):441–449.

Schmidhuber, J. and Tubiello, F.N. Global food security under climate change. *Proceedings of the National Academy of Science* 104, no. 50 (2007):19703–19708.

Schwind, K. Going local on a global scale: Rethinking food trade in an era of climate change, dumping and rural poverty. *Food First Backgrounder* 11, no.2 (2005):1–4.

Scoones, I. *Sustainable Rural Livelihoods: A Framework for Analysis*. Brighton, UK: University of Sussex Institute of Development Studies, 1998.

Scoones, I. and Thompson, J. *Beyond Farmer First: Rural Peoples Knowledge, Agricultural Research and Extension Practice*. London: Intermediate Technology Publications, 1994.

Scott, C. *The Moral Economy of the Peasant: Subsistence and Rebellion in Southeast Asia*. New Haven: Yale University Press, 1977.

———. *Weapons of the Weak: Everyday Forms of Peasant Resistance*. New Haven: Yale University Press, 1987.

Semple, H. *Agricultural Change and the Decline of Domestic Food Production in Guyana, 1960–1994*. Kingston: University of the West Indies Kingston, 1996.

Semple, H., Johnson, K. and Arjoonsingh, C. Agrochemicals usage and protective clothing among small farmers in Grenada and Jamaica. *Caribbean Geography* 14, no. 1(2005):15–30.

Sen, A. *Poverty and Famines: An Essay on Entitlement and Deprivation*. Oxford: Clarendon Press, 1981.

Shapiro, L. Hurricane climatic fluctuations. Part II: Relation to large-scale circulation. *Monthly Weather Review* 110 (1982):1007–1013.

Shapouri, S. and Rosen, S. *Food Security in Latin America and the Caribbean*. Washington, DC: United States department of agriculture, 2000.

Shepherd, A. Consolidating the lessons of 50 years of development. *Journal of International Development* 13 (2001):315–20.

Shipton, P. African famines and food security: Anthropological perspectives. *Annual Review of Anthropology* 19 (1990):353–394.

Shiva, V. *The Threat of the Globalization of Agriculture*. New Delhi: Research Foundation for Science, Technology and Natural Resource Policy, 1997.

Short, C. *Sustainable Food Security for all by 2020; Food Insecurity: A Symptom of Poverty*. London: Department for International Development (DFID), 2000. Retrieved from http://www.ifpri.org/2020conference/PDF/summary_short.pdf.

Sillitoe, P. The development of indigenous knowledge. *Current Anthropology*, 39 (1998):223–254.

———. Trust in development: Some implications of knowing in indigenous knowledge. *Journal of the Royal Anthropological Institute* 16 (2010):12–30.

Simon, H. A. Models of man: Social and rational. New York: Wiley 1957.

Slingo, J., Challinor, A., Hoskins, B. and Wheeler, T. Introduction: food crops in a changing climate. *Philosophical Transaction of the Royal Society of London B* 360 (2005):1983–1989.

Smit, B. and Skinner, M. Adaptation options in agriculture to climate change- a typology. *Mitigation and Adaptation Strategies for Global Change* 7(2002): 85–114.

Smit, B. and Wandel, J. Adaptation, adaptive capacity and vulnerability. *Global Environmental Change* 16(2006):282–292.

Smucker, T. and Wisner, B. Changing household responses to drought in Tharaka, Kenya: vulnerability persistence and challenge. *Disasters* 32, no.2 (2008):190–215.

Spence, B., Katada, T. and Clerveaux, V. *Experiences and Behaviour of Jamaican Residents in Relation in Hurricane Ivan*. Tokyo: Japan International Cooperation Agency, 2005.

Spitz, P. Food security, the right to food and the FAO. *FIAN-Magazine* no. 2 (2002).

Stamoulis, K. and Zezza, A. A conceptual framework for national agricultural, rural development, and food security strategies and policies. ESA Working Paper, No. 03–17, November, 2003. Agriculture and Economic Development Division, FAO, Rome. Available at www.fao.org/documents/show_cdr.asp?url_file=/docrep/ 007/ae050e/ac050e00.htm.

Stuart, T. *Waste: Uncovering the Global Food Scandal*. London: W.W. Norton Co., 2009.

Swanson, K. Impact of scaling behavior on tropical cyclone intensities [Electronic Version 2007]. *Geophysical Research Letters*, 34. Retrieved January 8, 2008 from http://meteo.lcd.lu/globalwarming/Swanson/Impact_of_scaling_TC_2007.pdf.

Talukder, A., Kiess, L., Huq, N., de Pee, S., Danton-Hill, I. and Bloem, M. increasing the production of vitamin A-rich fruits and vegetables. Lessons learned in taking the Bangladesh homegarden programme to a national scale. *Food and Nutrition Bulletin* 21, no. 2 (2000):165–172.

Tapper, D. and Font, X. *Tourism Supply Chains: Report on a Desk Research Project for the Travel Foundation*. Leeds: Leeds Metropolitan University, 2004.

Tartaglione, C., Smith, S. and O'Brien, J. *ENSO Impact on Hurricane Landfall Probabilities for the Caribbean*. Tallahassee, FL: Center for Ocean–Atmospheric Prediction Studies, The Florida State University, 2003.

Taylor, M., Enfield, D., Chen, A. The influence ofthe tropical Atlantic vs. the tropical Pacific on Caribbean rainfall. *Journal of Geophysical Resources*, 107, no.C9(2002):3127.

Thomas-Hope, E. and Jardine-Comrie, A. Caribbean agriculture in the new global environment. In Baker, G. ed. *No Island is an Island: The Impact of Globalization on the Commonwealth Caribbean*, London: Chatam House, 2007, 19–43.

Thomasson, D.A. Montserrat kitchen gardens: Social functions and development potential. *Caribbean Geography* 5, no.1 (1994):20–31.

Thompson, S. Food Crisis-Tufton Says Import Bill Out of Control-Urges Nation to Increase agricultural Production. *Daily Gleaner*, April 17, 2008.

tho Seeth, H., Chachnov, S. and Surinov, A. Russian Poverty: Muddling Through Economic Transition with Garden Plots. World Development 29, no.9 (1998):1611–1623.

Thrupp, L.A. Building legitimacy of indigenous knowledge: empowerment for third world people, or 'scientized packages' to be sold by development agencies. Farmers and Agricultural Research: Complementary Methods IDS Workshop 1987, University of Sussex.

Timms, B.F. Caribbean agriculture-tourism linkages in a neoliberal world: Problems and prospects for St Lucia. *International Development Planning Review* 28(2006):35–56.

———. Development theory and domestic agriculture in the Caribbean: recurring crises and missed opportunities. *Caribbean Geography* 15, no.2 (2008): 101–117.

Torres, R. Linkages between tourism and agriculture in Mexico. *Annals of Tourism Research*, 30, no. 3(2003):546–566.

Torres, R., Momsen, J. and Neimeier, D. A. Cuba's Farmers' Markets in the Special Period, in Besson, J. and J. Momsen eds. *Caribbean Land and Development Revisited*. New York: Palgrave Macmillan, 2007, 53–66.

Trueblood, M. and Shapouri, S. Implications of trade liberalization on food security of low-income countries; issues in food security. *Agriculture Information Bulletin* 765, no.5 (2001).

Tubiello, F. and Fischer, G. Reducing climate change impacts on agriculture: global and regional effects of mitigation, 2000–2080. *Technological Forecasting & Social Change* 74(2007), 1030–1056.

UN-HABITAT. *The State of the World's Cities: Globalization and Urban Culture.* Nairobi, Kenya: UN-HABITAT, 2004.

United Nations Development Program. *Urban Agriculture: Food, Jobs and Sustainable Cities.* UNDP Publication Series for habitat 2, Volume 1. New York: UNDP, 1996. United Nations Development Program. The Millenium Project 2005. *Halving Hunger: It Can Be Done.* Final report on the Task Force on Hunger, 2005. New York: Earth Institute at Columbia University. Available at http://www.unmilleniumproject.org/who/tf2docs.htm.

———. *Caribbean Human Development Report 2012.* Estimating a CARICOM Human Development Index. Rome: UNDP, 2012.

United States Department of Agriculture. Farmers' Market Growth. 2006. Available at http://www.ams.usda.gov/AMSv1.0/farmersmarkets

Van Den Ban, A.W. and Hawkins, H.S. *Agricultural Extension.* Longman Scientific and Technical USA and John Wiley and Sons Inc. New York, 1988.

Vedwan, N. Culture, climate and the environment: Local knowledge and perception of climate change among Apple Growers in Northwestern India. *Journal of Ecological Anthropology* 10 (2006):4–18.

Verveer, M. *The vital role of women in Agriculture and Rural Development.* Remarks to FAO. June 2011. Available at http://www.state.gov/s/gwi/rls/rem/2011/167899.htm.

Via Campesina. Food sovereignty: A future without hunger. 1996. Retrieved from http://www.viacampesina.org/imprimer.php3?id_article=38.pdf.

———. What is Food Sovereignty? 2003. Retrieved from http://www.viacampesina.org/IMG/_article_PDF/article_216.pdf.

Vincent, K. Uncertainty in adaptive capacity and the importance of scale. *Global Environmental Change* 17(2007):12–24.

Walelign, T. The fifth P7 summit: food sovereignty and democracy- Let the world feed itself. *GREEM/EFA International Relations Newsletter* no. 6 December, 2002.

Watts, D. *The West Indies: Patterns of Development, Culture and Environmental Change Since 1492.* Cambridge: Cambridge University Press, 1987.

Watts, M. *Silent Violence: Food, Famine and Peasantry in Northern Nigeria.* Berkeley: University of California Press, 1983.

Weis, T. Restructuring and redundancy: The impacts and illogic of neoliberal agricultural reforms in Jamaica. *Journal of Agrarian Change* 4, no.4 (2004):461–491.

———. Small farming and radical re-imaginations in the Caribbean today. *Race and Class* 49, no. 2 (2007):112–117.

Wiggins, S. Can the smallholder model deliver poverty reduction and food security for a rapidly growing population in Africa? Future Agricultures Consortium Working Paper No. 8.

Wigley, G.N. Constraints on soil conservation in the Pindars River and Two Meetings watersheds, Jamaica. Unpublished MA thesis, McGill University, 1988.

Williams, E. *From Columbus to Castro: The history of the Caribbean 1492–1969.* New York: Random House, Inc. 1970.

Windfuhr, M. Food security, food sovereignty, right to food: Competing or complementary approaches to fight hunger and malnutrition? Hungry for what is right. *FIAN-Magazine* no. 1, 2002.

———. *Food Sovereignty and the Right to Adequate Food.* Heidelburg, Germany: FIAN-International, 2003.

Windfuhr, M. and Jonsen, J. *Food Sovereignty: Towards Democracy in Localized Food Systems.* Rugby, UK: ITDG Publishing, 2005.

Witter, M. and Beckford, G. *Small Garden, Bitter Weed: The Political Economy of Struggle and Change in Jamaica.* Morant Bay, Jamaica: Maroon Publishing House, 1980.

World Bank. *World Development Report.* New York: The International Bank 1993.

———. *Missing Food: The Case of Postharvest Grain Losses in Sub-Saharan Africa.* Washington DC: World Bank, 2010.

———. Working to Rid the World of Poverty. Available at http://data.worldbank. org/indicator/AG.LND.AGRI.ZS.

———. *World Development Report 2008.* World Development Indicators, 2006. Washington DC: World Bank, 2008. Available at http://devdata.worldbank.org/ wdi2006/contents/index2.htm.

World Food Program. *Emergency Food Security Handbook.* Second Edition, January, 2009. http://www.wfp.org/operations/emergency_needs/EFSA_section1. pdf.World Food Summit. *Rome Declaration on World Food Security and World Food Summit Plan of Action, 1996.* Rome.

WTO. *World Trade Organization Annual Report, 1997.* Retrieved March 7, 2008, from http://www.wto.org/english/news_e/pres97_e/pr85_e.htm.

Zawisza, S. and Pilarski, S. Opinion Leadership and information sources in agricultural innovation diffusion processes (on the basis of selected villages in the Kujawsko-Pomorskie Province in Poland). Electronic Journal of Polish Agricultural Universities, *Economics* 8 no.4 (2005). Available at http://www.ejpau.media. pl/volume8/issue4/art-28.html.

Ziervogel, G., Taylor, A., Thomalla, F., Takama, K. and Quinn, C. *Adapting to Climate, Water and Health Stresses: Insights from, Sekhukhune, South Africa.* Lilla Nygatan, Sweden: Stockholm Environment Institute (SEI), 2006.

Index

Printed in the United States
by Baker & Taylor Publisher Services